Applied Probability Models with Optimization Applications

SHELDON M. ROSS

DOVER PUBLICATIONS, INC., New York

TO MY PARENTS

Published in Canada by General Publishing Company, Ltd., 30 Lesmill Road, Don Mills, Toronto, Ontario.

Published in the United Kingdom by Constable and Company, Ltd., 3 The Lanchesters, 162-164 Fulham Palace Road, London W6 9ER.

This Dover edition, first published in 1992, is an unabridged, unaltered republication of the work first published by Holden-Day, Inc., San Francisco, 1970, in the Holden-Day Series in Management Science.

Manufactured in the United States of America
Dover Publications, Inc., 31 East 2nd Street, Mineola, N.Y. 11501

Library of Congress Cataloging-in-Publication Data

Ross, Sheldon M.
   Applied probability models with optimization applications / Sheldon M. Ross.
      p.   cm.
   "An unabridged, unaltered republication of the work first published by Holden-Day, Inc., San Francisco, 1970, in the Holden-Day series in management science"—T.p. verso.
   Includes bibliographical references.
   ISBN 0-486-67314-6 (pbk.)
   1. Probabilities. 2. Mathematical optimization. I. Title.
QA273.R8 1992
519.2—dc20   92-16013
                                                    CIP

PREFACE

This text developed from a series of courses in applied probability and optimization theory given by the author at the University of California at Berkeley. A prerequisite to its reading would be some familiarity with probability theory at the level of Volume I of Feller's well-known book.

Ideally, this book would be used as a text in a one-year course in applied probability models with optimization applications. However, because of its flexibility, it would also be suitable as a text in more specialized topics. For instance, Chapters 1 through 5 and Chapter 9 might be used in a one- or two-quarter course in applied probability or stochastic processes. Also, Chapters 6 through 9 might be used as a text for a course in sequential decision theory. It should also be mentioned that this book has been designed so as to include most of the stochastic models of operations research, and hence may be readily used as a text in this subject.

The approach I have attempted to employ in this book is to view models from a probabilistic point of view. For instance, wherever possible, proofs are based on probabilistic, rather than analytic, arguments. I feel that this approach quickly gives the student a feeling for what is essential in this subject.

I am indebted to many people for their part in making this book possible. In particular, I would like to thank M. Brown, D. Iglehart, C. Myles Lavinsky, G. J. Lieberman, S. Lippman, and I. R. Savage. Finally, I must acknowledge Gary Ross, without whose help this book could not have been possible.

Sheldon M. Ross
Berkeley, California
October 1969

CONTENTS

Contents

Contents

1

INTRODUCTION TO
STOCHASTIC PROCESSES

1.1. Random Variables and Probability Theory

In order to understand the theory of stochastic processes, it is necessary to have a firm grounding in the basic concepts of probability theory. As a result, we shall briefly review some of these concepts and at the same time establish some useful notation.

The *distribution function* $F(\ )$ of the random variable X is defined for any real number x by

$$F(x) = P\{X \leq x\}$$

A random variable X is said to be *discrete* if its set of possible values is countable. For discrete random variables, the probability mass function $p(x)$ is defined by

$$p(x) = P\{X = x\}$$

Clearly,

$$F(x) = \sum_{y \leq x} p(y)$$

A random variable is called *continuous* if there exists a function $f(x)$, called the *probability density function*, such that

$$P\{X \text{ is in } B\} = \int_B f(x)\, dx$$

for every Borel set B. Since $F(x) = \int_{-\infty}^{x} f(x)\, dx$, it follows that

$$f(x) = \frac{d}{dx} F(x)$$

1

The *expectation* or *mean* of the random variable X, denoted by EX, is defined by

$$EX = \int_{-\infty}^{\infty} x \, dF(x) = \begin{cases} \displaystyle\int_{-\infty}^{\infty} xf(x) \, dx & \text{if } X \text{ is continuous} \\ \displaystyle\sum_{x:\, p(x) > 0} xp(x) & \text{if } X \text{ is discrete} \end{cases} \tag{1}$$

provided the above integral exists.

Equation (1) also defines the expectation of any function of X, say $h(X)$. Since $h(X)$ is itself a random variable, it follows from (1) that

$$Eh(X) = \int_{-\infty}^{\infty} x \, dF_h(x)$$

where F_h is the distribution function of $h(X)$. However, it can be shown that this is identical to $\int_{-\infty}^{\infty} h(x) \, dF(x)$, i.e.,

$$Eh(X) = \int_{-\infty}^{\infty} h(x) \, dF(x) \tag{2}$$

The above equation is sometimes known as the *law of the unconscious statistician* [as statisticians have been accused of using the identity (2) without realizing that it is not a definition].

The *variance* of the random variable X is defined by

$$\text{Var } X = E(X - EX)^2$$
$$= EX^2 - (EX)^2$$

Jointly Distributed Random Variables

The *joint distribution* F of two random variables X and Y is defined by

$$F(x, y) = P\{X \le x, \, Y \le y\}$$

The distributions $F_X(x) = \lim_{y \to \infty} F(x, y)$ and $F_Y(y) = \lim_{x \to \infty} F(x, y)$ are called the *marginal distributions* of X and Y. X and Y are said to be *independent* if

$$F(x, y) = F_X(x)F_Y(y)$$

for all real x and y. It can be shown that X and Y are independent if and only if

$$E[g(X)h(Y)] = E[g(X)]E[h(Y)]$$

for all functions g and h for which the above expectations exist.

Two jointly distributed random variables X and Y are said to be *uncorrelated* if their covariance, defined by

$$\text{Cov}(X, Y) = E[(X - EX)(Y - EY)]$$
$$= EXY - EXEY$$

is zero. It follows that independent random variables are uncorrelated. However, the converse is not true. (Give an example.)

The random variables X and Y are said to be *jointly continuous* if there exists a function $f(x, y)$, called the *joint probability density function*, such that

$$P\{X \text{ is in } A, Y \text{ is in } B\} = \int_A \int_B f(x, y)\, dy\, dx$$

for every two Borel sets A and B.

The joint distribution of any collection $X_1, X_2, \ldots, X_n$ of random variables is defined by

$$F(x_1, \ldots, x_n) = P\{X_1 \leq x_1, \ldots, X_n \leq x_n\}$$

Furthermore, the n random variables are said to be independent if

$$F(x_1, \ldots, x_n) = F_{X_1}(x_1) \ldots F_{X_n}(x_n)$$

where

$$F_{X_i}(x_i) = \lim_{\substack{x_j \to \infty \\ j \neq i}} F(x_1, \ldots, x_n)$$

Characteristic Functions and Laplace Transforms

The *characteristic function* $\phi(\ )$ of X is defined, for any real number u, by

$$\phi(u) = E[e^{iuX}]$$
$$= \int_{-\infty}^{\infty} e^{iux}\, dF(x)$$

A random variable always possesses a characteristic function (that is, the above integral always exists) and, in fact, it can be proven that there is a one-to-one correspondence between distribution functions and characteristic functions. This result is quite important as it enables us to characterize the probability law of a random variable by its characteristic function. (See Tables 1 and 2.)

A further useful result is that if $X_1, \ldots, X_n$ are independent, then the characteristic function of their sum $X_1 + \cdots + X_n$ is just the product of the individual characteristic functions. This result is quite useful, as it often

TABLE I

Discrete Probability Laws	Probability Mass Function $p(x)$	Characteristic Function	Mean	Variance
Poisson with parameter $\lambda > 0$	$p(x) = e^{-\lambda}\dfrac{\lambda^x}{x!}$ $x = 0, 1, 2, \ldots$	$e^{\lambda(e^{iu}-1)}$	λ	λ
Binomial with parameters n and p	$p(x) = \binom{n}{x} p^x(1-p)^{n-x}$ $x = 0, 1, \ldots, n$	$(pe^{iu} + q)^n$ where $q = 1 - p$	np	npq
Geometric $0 \le p \le 1$, $q = 1 - p$	$p(x) = pq^x$ $x = 0, 1, \ldots$	$\dfrac{p}{1 - qe^{iu}}$	$\dfrac{q}{p}$	$\dfrac{q}{p^2}$
Negative Binomial with parameters $r = 1, 2, \ldots$ and $p, 0 \le p \le 1$, $q = 1 - p$	$p(x) = \binom{r+x-1}{x} p^r q^x$ $x = 0, 1, 2, \ldots$	$\left(\dfrac{p}{1 - qe^{iu}}\right)^r$	$\dfrac{rq}{p}$	$\dfrac{rq}{p^2}$

TABLE 2

Continuous Probability Laws	Probability Density Function	Characteristic Function	Mean	Variance
Exponential $\lambda > 0$	$f(x) = \lambda e^{-\lambda x}$ $x \ge 0$	$\dfrac{\lambda}{\lambda - iu}$	$\dfrac{1}{\lambda}$	$\dfrac{1}{\lambda^2}$
Gamma $r > 0, \lambda > 0$	$f(x) = \dfrac{\lambda e^{-\lambda x}(\lambda x)^{r-1}}{\Gamma(r)}$ $x \ge 0$	$\left(\dfrac{\lambda}{\lambda - iu}\right)^r$	$\dfrac{r}{\lambda}$	$\dfrac{r}{\lambda^2}$
Uniform over $[a, b]$	$f(x) = \dfrac{1}{b - a}$ $a < x < b$	$\dfrac{e^{iub} - e^{iua}}{iu(b - a)}$	$\dfrac{a + b}{2}$	$\dfrac{(b - a)^2}{12}$
Normal μ, σ^2	$f(x) = \dfrac{1}{\sqrt{2\pi\sigma^2}} e^{-(x-\mu)^2/2\sigma^2}$ $-\infty < x < \infty$	$e^{iu\mu - u^2\sigma^2/2}$	μ	σ^2

enables us to determine the distribution of the sum of independent random variables by first calculating the characteristic function and then attempting to identify it.

EXAMPLE. Let X and Y be independent and identically distributed normal random variables having mean μ and variance σ^2. Then,

$$\phi_{X+Y}(u) = \phi_X(u)\phi_Y(u)$$

where the subscript indicates the random variable associated with the characteristic function. Hence (see Table 2),

$$\phi_{X+Y}(u) = (e^{i\mu u - \sigma^2 u^2/2})^2$$
$$= e^{i2\mu u - \sigma^2 u^2}$$

which is the characteristic function of a normal random variable having mean 2μ and variance $2\sigma^2$. Therefore, by the uniqueness of the characteristic function, this is the distribution of $X + Y$.

We may also define the joint characteristic of the random variables $X_1, \ldots, X_n$ by

$$\phi(u_1, u_2, \ldots, u_n) = E\left[\exp\left(i \sum_{i=1}^{n} u_i X_i\right)\right]$$

It may be proven that the joint characteristic function uniquely determines the joint distribution.

When dealing with random variables which only assume nonnegative values, it is sometimes more convenient to use *Laplace transforms* rather than characteristic functions. The Laplace transform of the distribution F (or, more precisely, of the random variable having distribution F) is defined by

$$\tilde{F}(s) = \int_0^\infty e^{-sx} \, dF(x)$$

This integral exists for a complex variable $s = a + bi$ where $a \geq 0$. As in the case of characteristic functions, the Laplace transform uniquely determines the distribution.

We may also define Laplace transforms for arbitrary functions in the following manner: The Laplace transform of the function g, denoted $\tilde{g}$, is defined by

$$\tilde{g}(s) = \int_0^\infty e^{-sx} \, dg(x)$$

provided the integral exists. It can be shown that $\tilde{g}$ determines g up to an additive constant.

Convolutions

If X and Y are independent random variables, with X having distribution F and Y having distribution G, then the distribution of $X + Y$ is given by $F * G$, where

$$(F * G)(t) = \iint_{X+Y \leq t} dF(x)\, dG(y) = \int_{-\infty}^{\infty} \int_{-\infty}^{t-y} dF(x)\, dG(y)$$

$$= \int_{-\infty}^{\infty} F(t - y)\, dG(y) = \int_{-\infty}^{\infty} G(t - x)\, dF(x)$$

$F * G$ is called the *convolution* of F and G. If $G = F$, then $F * F$ is denoted by F_2. Similarly, we denote by F_n the n-fold convolution of F with itself. That is,

$$F_n = F * \underbrace{(F * F * \cdots * F)}_{n-1}$$

It is easy to show that the characteristic function (Laplace transform) of a convolution is just the product of the characteristic functions (Laplace transforms).

Limit Theorems

Some of the most important results in probability theory are in the form of limit theorems. The two most important are:

Law of Large Numbers. If $X_1, X_2, \ldots$ are independent and identically distributed with mean μ, then with probability one,

$$\frac{X_1 + \cdots + X_n}{n} \to \mu \qquad \text{as } n \to \infty$$

Central Limit Theorem. If $X_1, X_2, \ldots$ are independent and identically distributed with mean μ and variance σ^2, then

$$\lim_{n \to \infty} P\left\{ \frac{X_1 + \cdots + X_n - n\mu}{\sigma\sqrt{n}} \leq a \right\} = \int_{-\infty}^{a} \frac{1}{\sqrt{2\pi}} e^{-x^2/2}\, dx$$

1.2. Conditional Expectation

If X and Y are discrete random variables, then the conditional probability mass function of Y, given X, is defined, for all x such that $P\{X = x\} > 0$, by

$$p_{Y|X}(y \mid x) = P\{Y = y \mid X = x\} = \frac{P\{Y = y, X = x\}}{P\{X = x\}}$$

Similarly, the conditional distribution function $F_{Y|X}(y|x)$ of Y, given X, is defined for all x such that $P\{X = x\} > 0$, by

$$F_{Y|X}(y|x) = P\{Y \le y|X = x\} = \sum_{y' \le y} p_{Y|X}(y'|x)$$

The conditional expectation of Y, given X, is defined for all x such that $P\{X = x\} > 0$, by

$$E[Y|X = x] = \int_{-\infty}^{\infty} y \, dF_{Y|X}(y|x) = \sum_{y} y p_{Y|X}(y|x)$$

If X and Y have a joint probability density function $f_{X,Y}(x, y)$, the conditional probability density function of Y, given X, is defined for all x such that $f_X(x) > 0$ by

$$f_{Y|X}(y|x) = \frac{f_{X,Y}(x, y)}{f_X(x)}$$

and the conditional probability distribution function of Y, given X, by

$$F_{Y|X}(y|x) = P\{Y \le y|X = x\} = \int_{-\infty}^{y} f_{Y|X}(y|x) \, dy$$

The conditional expectation of Y, given X, is defined for all x such that $f_X(x) > 0$, by

$$E[Y|X = x] = \int_{-\infty}^{\infty} y \, dF_{Y|X}(y|x) = \int_{-\infty}^{\infty} y f_{Y|X}(y|x) \, dy$$

Let us denote by $E[Y|X]$ the function of X whose value at $X = x$ is $E[Y|X = x]$. An extremely important property of conditional expectation is that for all random variables X and Y,

$$EY = E[E[Y|X]] = \int_{-\infty}^{\infty} E[Y|X = x] \, dF(x) \tag{3}$$

provided the relevant expectations exist. Hence, (3) states that the expectation of Y may be obtained by first conditioning on X (to obtain $E[Y|X]$) and then taking the expectation (with respect to X) of this quantity.

EXAMPLE. Suppose that the number of accidents occurring in a factory in a month is a random variable with distribution F. Suppose also that the number of workmen injured in each accident are independent and have a common distribution G. What is the expected number of workmen injured each month?

Let N denote the number of accidents which occur, and let $X_1, \ldots, X_N$ denote the number of workers injured in each accident. Then, the total number of workers injured is $\sum_{i=1}^{N} X_i$. Now

$$E \sum_{i=1}^{N} X_i = E\left\{ E\left[\sum_{i=1}^{N} X_i \mid N \right] \right\} \tag{4}$$

However,

$$E\left[\sum_{i=1}^{N} X_i \mid N = n \right] = E\left[\sum_{i=1}^{n} X_i \mid N = n \right]$$

$$= E \sum_{i=1}^{n} X_i$$

$$= n E X_1$$

Thus, from (4),

$$E \sum_{i=1}^{N} X_i = E[N E X_1] = E N \cdot E X_1$$

Example. A prisoner is placed in a cell containing three doors. The first door leads immediately to freedom. The second door leads into a tunnel which returns him to the cell after one day's travel. The third door leads to a similar tunnel which returns him to his cell after three days. Assuming that the prisoner is at all times equally likely to choose any one of the doors, what is the expected length of time until the prisoner reaches freedom?

Let Y denote the time until the prisoner reaches freedom, and let X denote the door that he initially chooses. We first note that

$$E[Y \mid X = 1] = 0$$
$$E[Y \mid X = 2] = 1 + EY \tag{5}$$
$$E[Y \mid X = 3] = 3 + EY$$

To see why this is so, consider $E[Y \mid X = 2]$, and reason as follows: If the prisoner chooses the second door, then he spends one day in the tunnel and then returns to his cell. But once he returns to his cell the problem is as before, and hence his expected time until freedom from that moment on is just EY. Hence, $E[Y \mid X = 2] = 1 + EY$. Therefore, from (5) we obtain

$$EY = \tfrac{1}{3}[0 + 1 + EY + 3 + EY]$$

or

$$EY = 4$$

Functional Equations and Lack of Memory of the Exponential Distribution

The following two functional equations occur quite frequently in the theory of applied probability:

$$f(s + t) = f(s) \cdot f(t) \qquad \text{for all } s, t \geq 0 \tag{6}$$

$$f(s + t) = f(s) + f(t) \qquad \text{for all } s, t \geq 0 \tag{7}$$

It turns out that the only (measurable) solution to these functional equations are of the respective forms

$$f(t) = e^{-\lambda t}$$

and

$$f(t) = ct$$

We shall now use (6) to prove that the exponential is the unique distribution without memory.

A random variable X is said to be without memory, or *memoryless*, if

$$P\{X > s + t \mid X > t\} = P\{X > s\} \qquad \text{for all } s, t \geq 0 \tag{8}$$

If we think of X as being the lifetime of some instrument, then (8) states that the probability that the instrument lives for at least $s + t$ hours given that it has survived t hours is the same as the initial probability that it lives for at least s hours. That is, the instrument does not deteriorate.

Suppose now that X is memoryless, and let $\bar{F}(x) = P\{X > x\}$. Now from (8) we obtain

$$\frac{P\{X > s + t, X > t\}}{P\{X > t\}} = P\{X > s\}$$

or

$$\bar{F}(s + t) = \bar{F}(s) \cdot \bar{F}(t)$$

implying that

$$\bar{F}(t) = e^{-\lambda t}$$

which is the distribution of the exponential random variable. Also, by considering the argument in reverse, it follows that the exponential distribution is memoryless.

1.3. Stochastic Processes

A *stochastic process* $\{X(t), t \in T\}$ is a family of random variables. That is, for each t contained in the index set T, $X(t)$, is a random variable. The variable t is often interpreted as time, and hence $X(t)$ represents the *state* of the process at time t. For instance, $X(t)$ may represent the amount of inventory in a retail store at time t or the number of people in a bank at time t or the position of a particle at time t, etc.

The set T is called the *index set* of the stochastic process. If T is a countable set, then the stochastic process is said to be a *discrete time* process. If T is an open or closed interval of the real line, then we say that the stochastic process is a *continuous time* process.

The set of possible values which the random variables $X(t)$, $t \in T$ may assume is called the *state space* of the process.

A continuous time stochastic process $\{X(t), t \in T\}$ is said to have *independent increments* if for all choices of $t_0 < t_1 < t_2 < \cdots < t_n$, the n random variables

$$X(t_1) - X(t_0), X(t_2) - X(t_1), \ldots, X(t_n) - X(t_{n-1})$$

are independent. The process is said to have *stationary independent increments* if in addition $X(t_2 + s) - X(t_1 + s)$ has the same distribution as $X(t_2) - X(t_1)$ for all $t_1, t_2 \in T$ and $s > 0$.

Examples of Stochastic Processes

1. The General Random Walk
    Let $Y_1, Y_2, \ldots$ be a sequence of independent and identically distributed random variables, and let $X_n = \sum_{i=1}^{n} Y_i$. The stochastic process

$$\{X_n, n = 0, 1, 2, \ldots\}$$

is called the *general random walk process*. If Y_i represents the number of items sold by a retail store during the ith week, then X_n would be the total number of items sold during the first n weeks.

2. The Wiener Process
    A stochastic process $\{X(t), t \geq 0\}$ is said to be a *Wiener process* if

    (i) $\{X(t), t \geq 0\}$ has stationary, independent increments
    (ii) for every $t > 0$, $X(t)$ is normally distributed with mean 0
    (iii) $X(0) = 0$.

The Wiener process will be extensively studied in Chapter 9.

Both the general random walk process and the Wiener process are examples of a class of stochastic processes known as *Markov processes*. A Markov process is a stochastic process with the property that given the value of $X(t)$, the probability of $X(s + t)$, where $s > 0$, is independent of the values of $X(u)$, $u < t$. That is, the conditional distribution of the future $X(s + t)$, given the present $X(t)$ and the past $X(u)$, $u < t$, is independent of the past. More formally, the process $\{X(t), t \in T\}$ is said to be a Markov process if

$$P\{X(t) \le x \mid X(t_1) = x_1, X(t_2) = x_2, \ldots, X(t_n) = x_n\}$$
$$= P\{X(t) \le x \mid X(t_n) = x_n\}$$

whenever $t_1 < t_2 < \cdots < t_n < t$.

Stochastic Processes in Queueing Systems

Queueing theory is the study of waiting line phenomena. A queue is generated when customers arrive at a station to receive service. If the customer arrives when all the servers are busy, then he must join the queue (i.e., wait in line).

We shall adopt the following notation in describing certain queueing situations. It is assumed that the successive interarrival times T_1, T_2, ... between customers are independent and identically distributed random variables having a distribution F. (That is, T_1 is the time of the first arrival, T_2 the time between the first and second arrival, etc.) Suppose also that the service times of successive customers are also independent and have a common distribution G. Finally, let s denote the number of servers. We use the notation $F/G/s$ to describe this queueing system. The following symbols are used to denote interarrival and service time distributions:

M (for memoryless) for the exponential distributions
G for a general distribution

Thus, $M/G/1$ denotes a queueing system with exponential interarrival times, an arbitrary service time distribution, and a single server. The system $G/G/1$ is the single server queueing system with an arbitrary interarrival and arbitrary service distribution. (The interarrival and service distribution in the $G/G/1$ system need not be the same.)

There are many stochastic processes associated with queueing systems. For instance, $X(t)$ may represent the number of customers in the system at t, or the number who have been serviced by time t. An example of a discrete time stochastic process is the process $\{W_n, n = 1, 2, \ldots\}$ where W_n represents the amount of time the nth customer spends in the system.

Problems

1. Show that $\text{Var}(X + Y) = \text{Var } X + \text{Var } Y + 2 \text{ Cov}(X, Y)$.

2. If X has characteristic function $\phi(u)$, show that

$$EX = \frac{1}{i} \frac{d}{du} \phi(u)|_{u=0},$$

and in general,

$$EX^n = \left(\frac{1}{i}\right)^n \frac{d^n}{du^n} \phi(u)|_{u=0}$$

whenever the moments exist.

3. If X and Y are independent Poisson random variables, find the distribution of $X + Y$.

4. Let X and Y be Poisson with respective means λ_1 and λ_2. Show that the conditional distribution of X, given $X + Y$, is binomial.

5. Suppose X is distributed as a Poisson random variable with mean λ. The parameter λ is itself a random variable whose distribution law is exponential with mean $1/\mu$. Find the distribution of X.

6. An urn has n chips. Chips are drawn one at a time and then put back in the urn. Let N denote the number of drawings required until some chip is drawn more than once. Find the probability distribution of N.

7. A man with n keys wants to open his door. He tries the keys in a random manner. Let N be the number of trials required to open the door. Find EN and $\text{Var } N$ if (a) unsuccessful keys are eliminated from further selection, (b) if they are not.

8. The conditional variance of Y, given X, is defined by

$$\text{Var}(Y|X) = E[(Y - E(Y|X))^2 | X]$$

Show that $\text{Var } Y = E[\text{Var}(Y|X)] + \text{Var } E[Y|X]$.

9. Let N denote the number of customers arriving at a store in a given day. Suppose that the amounts spent by the customers are independent and have a distribution F. Find the mean and variance of the total amount of money spent in the store.

10. Show that every stochastic process with independent increments is a Markov process

11. Let $Y_{1n}, Y_{2,n}, \ldots, Y_{n,n}$ be n independent random variables with the identical uniform distribution on $(0, t)$. Let $Z_n = \min(Y_{1,n}, \ldots, Y_{n,n})$.

(a) Find $P\{Z_n > x\}$.

(b) Let t be a function of n such that $\lim_{n \to \infty} n/t = \lambda$.

Show that

$$\lim_{n \to \infty} P\{Z_n > x\} = e^{-\lambda x}.$$

12. Show that the only continuous solutions of the functional Equations (6) and (7) are respectively $f(t) = e^{-\lambda t}$ and $f(t) = ct$.

13. Show for the Wiener process that $\text{Var } X(t) = \sigma^2 t$ for some $\sigma^2 > 0$.

2

THE POISSON PROCESS

2.1. Introduction and Definitions

A stochastic process $\{N(t), t \geq 0\}$ is said to be a *counting process* if $N(t)$ represents the total number of events which have occurred up to time t. A particularly important counting process is the Poisson process.

Definition 2.1

The counting process $\{N(t), t \geq 0\}$ is said to be a *Poisson Process* if

(i) $N(0) = 0$
(ii) $\{N(t), t \geq 0\}$ has independent increments
(iii) The number of events in any interval of length t has a Poisson distribution with mean λt. That is, for all $s, t \geq 0$,

$$P\{N(t + s) - N(s) = n\} = e^{-\lambda t} \frac{(\lambda t)^n}{n!}, \qquad n \geq 0$$

From (iii), it follows that

$$E[N(t)] = \lambda t$$

and λ is called the *rate of the process*.

It is also possible to give a more physical definition of the Poisson process. We first introduce the following concept:

NOTATION. A function f is said to be $o(t)$ if

$$\lim_{t \to 0} \frac{f(t)}{t} = 0$$

13

Definition 2.2

$\{N(t), t \geq 0\}$ is a Poisson process if

(i) $N(0) = 0$
(ii) $\{N(t), t \geq 0\}$ has stationary, independent increments
(iii) $P\{N(t) \geq 2\} = o(t)$
(iv) $P\{N(t) = 1\} = \lambda t + o(t)$

Theorem 2.1

Definitions 2.1 and 2.2 are equivalent.

PROOF. To show that Definition 2.2 implies Definition 2.1, let

$$P_n(t) = P\{N(t) = n\}$$

We derive a differential equation for $P_0(t)$ in the following manner:

$$\begin{aligned}
P_0(t + h) &= P\{N(t + h) = 0\} \\
&= P\{N(t) = 0, N(t + h) - N(t) = 0\} \\
&= P\{N(t) = 0\} \cdot P\{N(t + h) - N(t) = 0\} \\
&= P_0(t) \cdot P_0(h)
\end{aligned}$$

where Assumption (ii) has been used in obtaining the final two equations. Hence,

$$\frac{P_0(t + h) - P_0(t)}{h} = P_0(t) \frac{(P_0(h) - 1)}{h}$$

Now, letting $h \to 0$ and using the fact that (iii) and (iv) imply that $P_0(h) = 1 - \lambda h + o(h)$, we obtain

$$P_0'(t) = -\lambda P_0(t)$$

or equivalently,

$$\log P_0(t) = -\lambda t + c$$

or

$$P_0(t) = ce^{-\lambda t}$$

Since $P_0(0) = 1$, we arrive at

$$P_0(t) = e^{-\lambda t} \tag{1}$$

Similarly, for $n > 0$,

$$
\begin{aligned}
P_n(t + h) &= P\{N(t + h) = n\} \\
&= P\{N(t) = n, N(t + h) - N(t) = 0\} \\
&\quad + P\{N(t) = n - 1, N(t + h) - N(t) = 1\} \\
&\quad + \sum_{k=2}^{n} P\{N(t) = n - k, N(t + h) - N(t) = k\}
\end{aligned}
$$

However, by (iii), the last term in the above is $o(h)$; hence, by using (ii), we obtain

$$
\begin{aligned}
P_n(t + h) &= P_n(t)P_0(h) + P_{n-1}(t)P_1(h) + o(h) \\
&= (1 - \lambda h)P_n(t) + \lambda h P_{n-1}(t) + o(h)
\end{aligned}
$$

Thus,

$$
\frac{P_n(t + h) - P_n(t)}{h} = -\lambda P_n(t) + \lambda P_{n-1}(t) + \frac{o(h)}{h}
$$

and letting $h \to 0$ yields

$$
P_n'(t) = -\lambda P_n(t) + \lambda P_{n-1}(t)
$$

or equivalently,

$$
e^{\lambda t}[P_n'(t) + \lambda P_n(t)] = \lambda e^{\lambda t} P_{n-1}(t)
$$

Hence,

$$
\frac{d}{dt}(e^{\lambda t}P_n(t)) = \lambda e^{\lambda t} P_{n-1}(t) \tag{2}
$$

Now by (1), we have

$$
\frac{d}{dt}(e^{t\lambda}P_1(t)) = \lambda
$$

or

$$
P_1(t) = (\lambda t + c)e^{-\lambda t}
$$

which, since $P_1(0) = 0$, yields

$$
P_1(t) = \lambda t e^{-\lambda t}
$$

To show that $P_n(t) = e^{-\lambda t}(\lambda t)^n/n!$, we first assume it for $n - 1$. Then by (2),

$$
\frac{d}{dt}(e^{\lambda t}P_n(t)) = \frac{\lambda^n t^{n-1}}{(n-1)!}
$$

or

$$
e^{\lambda t}P_n(t) = \frac{(\lambda t)^n}{n!} + c
$$

which implies the result (since $P_n(0) = 0$). This proves that Definition 2.2 implies Definition 2.1. We shall leave it for the reader to prove the reverse.

2.2. Interarrival and Waiting Time Distributions

Let X_1 denote the time from 0 to the first event, and for $n > 1$, let X_n denote the time from the $(n-1)$st to the nth event. The sequence

$$\{X_n, n = 1, 2, \ldots\}$$

is called the *sequence of interarrival times.*

In order to determine the distribution of the X_n, we first note that

$$P\{X_1 > t\} = P\{N(t) = 0\} = e^{-\lambda t}$$

Hence, X_1 has an exponential distribution with mean $1/\lambda$. Now,

$$P\{X_2 > t\} = E[P\{X_2 > t \mid X_1\}] \tag{3}$$

However,

$$
\begin{aligned}
P\{X_2 > t \mid X_1 = s\} &= P\{0 \text{ events in } (s, s + t] \mid X_1 = s\} \\
&= P\{0 \text{ events in } (s, s + t]\} \\
&= e^{-\lambda t}
\end{aligned}
$$

where the last two equations followed from independent and stationary increments. Therefore, from (3) we conclude that X_2 is also an exponential random variable with mean $1/\lambda$, and furthermore, that X_2 is independent of X_1. Repeating the same argument yields

Proposition 2.2

$X_n, n = 1, 2, \ldots$ are independent identically distributed exponential random variables having mean $1/\lambda$.

REMARK. Proposition 2.2 should not surprise us. The assumption of stationary and independent increments is basically equivalent to asserting that, at any point in time, the process *probabilistically* restarts itself. That is, the process from this point on is independent of all that has previously occurred (by independent increments), and also has the same distribution as the original process (by stationary increments). In other words, the process has no *memory*, and hence exponential interarrival times are to be expected.

Another quantity of interest is S_n, the arrival time of the nth event, also called the *waiting time* until the nth event. It is easily seen that

$$S_n = \sum_{i=1}^{n} X_i, \qquad n \geq 1$$

and hence from Proposition 2.2 one may routinely show, by using characteristic functions, that S_n has a gamma distribution with parameters n and λ. That is, the probability density function of S_n is given by

$$f_{S_n}(x) = \lambda e^{-\lambda x} \frac{(\lambda x)^{n-1}}{(n-1)!} \qquad x \geq 0 \tag{4}$$

Equation (4) may also be derived by noting that

$$N(t) \geq n \Leftrightarrow S_n \leq t$$

Hence,

$$F_{S_n}(t) = \sum_{j=n}^{\infty} e^{-\lambda t} \frac{(\lambda t)^j}{j!}$$

which yields (4) upon differentiation.

2.3. Conditional Distribution of the Arrival Times

Suppose we are told that exactly one event has taken place in the interval $[0, t]$, and we are asked to determine the distribution of the time at which the event occurred. Now, because of the assumption of stationary and independent increments, it seems reasonable that each interval in $[0,t]$ of equal length should have the same probability of containing the event. In other words, the time of the event should have a uniform distribution over $[0,t]$.

This is easily checked since, for $s \leq t$,

$$P\{X_1 < s \mid N(t) = 1\} = \frac{P\{N(s) = 1, N(t) - N(s) = 0\}}{P\{N(t) = 1\}} = \frac{s}{t}$$

This result may be generalized, but before doing so we need to introduce the concept of order statistics.

Let $Y_1, Y_2, \ldots, Y_n$ be n random variables. We say that $Y_{(1)}, Y_{(2)}, \ldots, Y_{(n)}$ are the order statistics corresponding to $Y_1, Y_2, \ldots, Y_n$ if $Y_{(k)}$ is the kth smallest value among $Y_1, \ldots, Y_n$, $k = 1, 2, \ldots, n$. If the Y_i's are independent identically distributed continuous random variables with probability density function f, then the joint distribution of the order statistics is given by

$$f_{Y_{(1)}, \ldots, Y_{(n)}}(y_1, \ldots, y_n) = n! \, f(y_1) \ldots f(y_n) \qquad y_1 < y_2 \cdots < y_n$$

This is true as each of the $n!$ permutations of $Y_1, \ldots, Y_n$ leads to the same order statistics. We are now ready for the following useful theorem.

Theorem 2.3

Given that $N(t) = n$, the n arrival times $S_1, \ldots, S_n$ have the same distribution as the order statistics corresponding to n independent random variables uniformly distributed on the interval $(0, t)$.

PROOF. Suppose $0 < t_1 < t_2 < \cdots < t_n < t_{n+1} \equiv t$; and let h_i be small enough so that $t_i + h_i < t_{i+1}$, $i = 1, 2, \ldots, n$. Now,

$P\{t_i \leq S_i \leq t_i + h_i, \qquad i = 1, 2, \ldots, n \mid N(t) = n\}$

$$= \frac{\begin{array}{l} P\{\text{exactly 1 event in } [t_i, t_i + h_i], \\ \qquad i = 1, 2, \ldots, n, \text{ no events elsewhere in } [0, t]\} \end{array}}{P\{N(t) = n\}}$$

$$= \frac{\lambda h_1 e^{-\lambda h_1} \cdots \lambda h_n e^{-\lambda h_n} e^{-\lambda(t - h_1 - h_2 \cdots - h_n)}}{e^{-\lambda t} (\lambda t)^n / n!}$$

$$= \frac{n!}{t^n} h_1 \cdot h_2 \cdot \cdots \cdot h_n$$

Hence,

$$\frac{P\{t_i \leq S_i \leq t_i + h_i, \qquad i = 1, 2, \ldots, n \mid N(t) = n\}}{h_1 \cdot h_2 \cdot \cdots \cdot h_n} = \frac{n!}{t^n}$$

and by letting the $h_i \to 0$, we obtain

$$f_{S_1, S_2, \ldots, S_n}(t_1, t_2, \ldots, t_n \mid N(t) = n) = \frac{n!}{t^n} \qquad 0 < t_1 < \cdots < t_n < t$$

and the result is obtained.

REMARK. Intuitively, we usually say that under the condition that n events have occurred in $(0, t)$, the times $S_1, \ldots, S_n$ at which events occur, considered as unordered random variables, are distributed independently and uniformly in the interval $(0, t)$.

EXAMPLE. *An Infinite Server Poisson Queue.* Suppose that customers arrive at a service station in accordance with a Poisson process with rate λ. Upon arrival, the customer is immediately served by one of an infinite number of servers, and the service times are assumed to be independent with a common distribution G.

Let $X(t)$ denote the number of customers in the system at t. We shall determine the distribution of $X(t)$ by conditioning on $N(t)$, the total number of customers who have arrived by t. By conditioning, we obtain

$$P\{X(t) = j\} = \sum_{n=0}^{\infty} P\{X(t) = j \mid N(t) = n\} e^{-\lambda t} \frac{(\lambda t)^n}{n!} \tag{5}$$

Now the probability that a customer who arrives at time x will still be present at t is $1 - G(t - x)$. Hence, given that $N(t) = n$, it follows by Theorem 2.3

that the probability of an arbitrary one of these customers still being present at t is given by

$$p = \int_0^t (1 - G(t - x)) \frac{dx}{t} = \int_0^t (1 - G(x)) \frac{dx}{t}$$

independently of the others. Hence,

$$P\{X(t) = j \mid N(t) = n\} = \begin{cases} \binom{n}{j} p^j (1 - p)^{n-j} & j = 0, 1, \ldots, n \\ 0 & j > n \end{cases}$$

and thus by (5),

$$P\{X(t) = j\} = \sum_{n=j}^{\infty} \binom{n}{j} p^j (1 - p)^{n-j} e^{-\lambda t} \frac{(\lambda t)^n}{n!}$$

$$= e^{-\lambda t} \frac{(\lambda t p)^j}{j!} \sum_{n=j}^{\infty} \frac{(\lambda t (1 - p))^{n-j}}{(n - j)!}$$

$$= e^{-\lambda t p} \frac{(\lambda t p)^j}{j!}$$

where

$$p = \int_0^t \frac{(1 - G(x))}{t} \, dx.$$

Or, in other words, $X(t)$ has a Poisson distribution with mean $\lambda \int_0^t (1 - G(x)) dx$.

EXAMPLE. *An Electronic Counter.* Suppose that electrical pulses having random amplitudes arrive at a counter in accordance with a Poisson process with rate λ. The amplitude of the pulses is assumed to decrease an exponential rate. That is, we suppose that if the pulse has an amplitude of A units upon arrival, then its amplitude at a time t units later will be $A e^{-\alpha t}$. We further suppose that the initial amplitudes of the pulses are independent and have a common distribution F.

Let $S_1, S_2, \ldots$ be the arrival times of the pulses and let $A_1, A_2, \ldots$ be their respective amplitudes. Then

$$A(t) = \sum_{n=1}^{N(t)} A_n e^{-\alpha(t - S_n)}$$

represents the total amplitude at time t. We shall determine the distribution of $A(t)$ by calculating its characteristic function.

Let

$$\phi_{A(t)}(u) = E[e^{iuA(t)}]$$

$$= \sum_{n=0}^{\infty} E[e^{iuA(t)} \mid N(t) = n] e^{-\lambda t} \frac{(\lambda t)^n}{n!}$$

Now, conditioned on $N(t) = n$, the unordered arrival times $(S_1, \ldots, S_n)$ are distributed as independent uniform $(0, t)$ random variables. Hence, given $N(t) = n$, $A(t)$ has the same distribution as $\sum_{j=1}^{n} A_j e^{-\alpha(t-Y_j)}$, where $Y_j, j = 1, \ldots, n$ are independent and uniformly distributed on $(0, t)$. Thus,

$$E[e^{iuA(t)} \mid N(t) = n] = E\left[\exp\left\{iu \sum_{j=1}^{n} A_j e^{-\alpha(t-Y_j)}\right\}\right]$$

$$= E[\exp\{iuA_1 e^{-\alpha(t-Y_1)}\}]^n$$

Now,

$$E\{\exp[iuA_1 e^{-\alpha(t-Y_1)}] \mid Y_1 = y\} = \phi_A[ue^{-\alpha(t-y)}]$$

where $\phi_A(u)$ is the characteristic function of A_1. Hence,

$$E[e^{iuA(t)} \mid N(t) = n] = \left\{\int_0^t \phi_A[ue^{-\alpha(t-y)}] \frac{dy}{t}\right\}^n$$

$$= \left\{\frac{1}{t} \int_0^t \phi_A(ue^{-\alpha y}) \, dy\right\}^n$$

and therefore,

$$\phi_{A(t)}(u) = \sum_{n=0}^{\infty} e^{-\lambda t} \frac{(\lambda t)^n}{n!} \left[\frac{1}{t} \int_0^t \phi_A(ue^{-\alpha y}) \, dy\right]^n$$

$$= \exp\left\{-\lambda \int_0^t [1 - \phi_A(ue^{-\alpha y})] \, dy\right\}$$

The moments of $A(t)$ may then be calculated by differentiation. For example,

$$E[A(t)] = \phi'_{A(t)}(0)/i$$

$$= \lambda \int_0^t \frac{\phi'_A(0)}{i} e^{-\alpha y} \, dy$$

$$= \frac{\lambda E A_1}{\alpha} (1 - e^{-\alpha t})$$

AN OPTIMIZATION EXAMPLE. Suppose that items arrive at a processing plant in accordance with a Poisson process with rate λ. At a fixed time T, all items are dispatched from the system. The problem is to choose an intermediate time, $t \in (0, T)$, at which all items in the system are dispatched, so as to minimize the total wait of all items.

If we dispatch at time t, $0 < t < T$, then the expected wait of all items will be

$$\frac{\lambda t^2}{2} + \frac{\lambda (T - t)^2}{2} \tag{6}$$

To see why (6) is true, we reason as follows: The expected number of arrivals in $(0, t)$ is λt, and each arrival is uniformly distributed on $(0, t)$, and hence has expected wait $t/2$. Thus, the expected total wait of items arriving in $(0, t)$ is $\lambda t^2/2$. Similar reasoning holds for arrivals in (t, T), and (6) follows.

Hence, we see from (6) that the constant dispatch time minimizing the expected total wait of all items is $T/2$, and the minimal expected waiting time is $\lambda T^2/4$.

In fact, we can say more than this. It turns out that not only does $T/2$ minimize the expected wait, it also maximizes the probability that the wait is less than y, for every $y > 0$. To see this, let t $(0 < t < T)$ be any intermediate dispatch. Then the total waiting time is $\sum_{i=1}^{N(T)} W_t(i)$, where $N(T)$ denotes the number of items arriving in $[0, T]$, and

$$W_t(i) = \begin{cases} t - S_i & \text{if } S_i \le t \\ T - S_i & \text{if } S_i > t \end{cases}$$

where S_i, $i = 1, \ldots, N(T)$ are the times at which items arrive (see Figure 2.1).

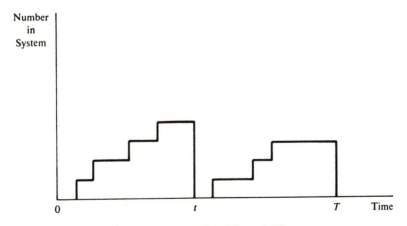

Figure 2.1. Intermediate Dispatch Time t

Now, conditional on the event that $N(T) = n$, the (unordered) points $S_1, S_2, \ldots, S_n$ are independent and identically distributed (*iid*) as uniform random variables on $(0, T)$; thus, $\sum_{i=1}^{N(T)} W_t(i)$ has the same distribution as $\sum_{i=1}^{N} Z_t(i)$, where N is distributed as a Poisson random variable with mean λT, and the $Z_t(i)$ are *iid* independent of N, having

$$P\{Z_t(i) < y\} = \begin{cases} \dfrac{\min(y, t) + \min(y, T - t)}{T} & y \leq T \\[2mm] 1 & y > T \end{cases} \tag{7}$$

The reasoning behind (7) is as follows: To have a wait of less than y, the item must arrive either between

$$\begin{cases} (t - y, t) & \text{if } t - y > 0 \\ (0, t) & \text{if } t - y < 0 \end{cases}$$

or between

$$\begin{cases} (T - y, T) & \text{if } T - y > t \\ (t, T) & \text{if } T - y < t \end{cases}$$

Hence, it would have to arrive in one of two disjoint intervals having lengths of $\min(y, t)$ and $\min(y, T - t)$. The result (7) then follows, since the arrival time of the item is uniformly distributed on $(0, T)$.

From (7), it follows that for fixed y, $P\{Z_t(i) < y\}$ is maximized by $t = T/2$. Since the distribution of N is independent of t and of the $Z_t(i)$'s, it follows that

$$P\left\{ \sum_{i=1}^{N} Z_t(i) < y \right\} = P\left\{ \sum_{i=1}^{N(T)} W_t(i) < y \right\}$$

is maximized by $t = T/2$.

Thus, not only does $T/2$ minimize the expected total waiting time, but it also possesses the stronger property of maximizing the probability that the total wait is less than y for every $y > 0$. (See Problem 20.)

REMARK. Theorem 2.3 may also be used to test the hypothesis that a given counting process is a Poisson process. This may be done by observing the process for a fixed time t. If in this time period we observe n arrivals, then if the process is Poisson, the unordered arrival times would be independent and uniformly distributed on $(0, t)$. Hence, we may test if the process is Poisson by testing the hypothesis that the n arrival times come from a uniform $(0, t)$ population. This may be done by standard statistical procedures (such as the Kolmogorov-Smirnov test).

2.4. Compound and Nonhomogeneous Poisson Processes

A stochastic process $\{X(t), t \geq 0\}$ is said to be a *compound Poisson process* if it can be represented, for $t \geq 0$, by

$$X(t) = \sum_{i=1}^{N(t)} Y_i$$

where $\{N(t), t \geq 0\}$ is a Poisson process, and $\{Y_n, n = 1, 2, \ldots\}$ is a family of independent and identically distributed random variables. The process $\{N(t), t \geq 0\}$ and the sequence $\{Y_n\}$ are assumed to be independent.

As an example of a compound Poisson process, suppose customers enter a store at a Poisson rate λ. Suppose, also, that the number of items purchased by each customer forms a set of independent and identically distributed random variables. Then, letting $X(t)$ denote the total number of items purchased by t, we see that $\{X(t), t \geq 0\}$ is a compound Poisson process.

We now calculate the characteristic function of $X(t)$. Let

$$\phi_t(u) = E[e^{iuX(t)}] \tag{8}$$

By conditioning on $N(t)$, we obtain

$$\phi_t(u) = \sum_{n=0}^{\infty} E[e^{iuX(t)} \mid N(t) = n] e^{-\lambda t} \frac{(\lambda t)^n}{n!}$$

However

$$E[e^{iuX(t)} \mid N(t) = n] = E\{\exp[iu(Y_1 + \cdots + Y_n)] \mid N(t) = n\} \tag{9}$$

$$= E\{\exp[iu(Y_1 + \cdots + Y_n)]\}$$

$$= \{E[\exp(iu Y_1)]\}^n \tag{10}$$

where (9) follows from the independence of $\{Y_1, Y_2, \ldots\}$ and $N(t)$, and (10) from the independence of the Y_i's. Hence, letting

$$\phi_y(u) = E[e^{iuY_1}]$$

we have from (8) and (10) that

$$\phi_t(u) = \sum_{n=0}^{\infty} \phi_y^n(u) e^{-\lambda t} \frac{(\lambda t)^n}{n!}$$

$$= e^{\lambda t [\phi_y(u) - 1]}$$

By differentiation of the above, we obtain

$$E[X(t)] = \phi_t'(0)/i = \lambda t EY \tag{11}$$

and

$$\text{Var}[X(t)] = -\phi_t''(0) - [EX(t)]^2 = \lambda t EY^2 \tag{12}$$

Another generalization of the Poisson process is attained by allowing the rate at time t to be a function of t. More formally, we have the following definition.

Definition 2.3

$[N(t), t \geq 0]$ is said to be a *nonhomogeneous Poisson process* with intensity function $\lambda(t)$ if

(i) $N(0) = 0$
(ii) $[N(t), t \geq 0]$ has independent increments
(iii) $P[2$ or more events in $(t, t + h)] = o(h)$
(iv) $P[\text{exactly 1 event in } (t, t + h)] = \lambda(t)h + o(h)$.

If we let

$$m(t) = \int_0^t \lambda(s) \, ds$$

then it can be shown that

$$P[N(t) = n] = e^{-m(t)} \frac{[m(t)]^n}{n!}, \qquad n \geq 0 \tag{13}$$

Or, in other words, $N(t)$ has a Poisson distribution with mean $m(t)$. $m(t)$ is said to be the *mean value function* of the process.

2.5. Stationary Point Processes

In this section, we shall study counting processes which have stationary, but not necessarily independent, increments. Such a process $\{N(t), t \geq 0\}$ is usually called a *stationary point process*.

As before, let $P_n(t) = P[N(t) = n]$.

Theorem 2.4

For any stationary point process, excluding the trivial case $P_0(t) = 1$ for all t,

$$\lim_{t \to 0} \frac{1 - P_0(t)}{t} = \lambda > 0 \tag{14}$$

where $\lambda = \infty$ is not excluded.

PROOF. Let $f(t) = 1 - P_0(t)$ and note that $f(t)$ is nonnegative and non-decreasing. Furthermore, since an event in $(0, s + t]$ implies an event in one of the intervals $(0, s]$, $(s, s + t]$ we obtain $f(s + t) \le f(s) + f(t)$. Hence, by induction,

$$f(t) \le nf(t/n) \qquad \text{for all } n = 1, 2, \ldots \tag{15}$$

Thus, letting a be such that $f(a) > 0$, we have

$$\frac{f(a)}{a} \le \frac{f(a/n)}{a/n} \qquad \text{all } n = 1, 2, \ldots \tag{16}$$

Now define $\lambda = \lim\sup_{t \to 0} f(t)/t$. By (16) we obtain

$$\lambda \ge \frac{f(a)}{a} > 0$$

To show that $\lambda = \lim_{t \to 0} f(t)/t$, we consider two cases. First, suppose that $\lambda < \infty$. In this case, fix $\varepsilon > 0$, and let $s > 0$ be such that $f(s)/s > \lambda - \varepsilon$. Now, for any $t \in (0, s)$ there is an integer n such that

$$\frac{s}{n} \le t \le \frac{s}{n-1}$$

From (15) and the monotonicity of $f(t)$, we obtain for all t in this interval,

$$f(t)/t \ge \frac{f(s/n)}{s/(n-1)} = \frac{n-1}{n} \frac{n}{s} f(s/n) \ge \frac{n-1}{n} \frac{f(s)}{s} \tag{17}$$

and hence,

$$f(t)/t > \frac{n-1}{n} (\lambda - \varepsilon)$$

Since ε is arbitrary and since $n \to \infty$ as $t \to 0$, it follows that $\lim_{t \to 0} f(t)/t = \lambda$.

Now assume $\lambda = \infty$. In this case, fix any large $A > 0$ and choose s such that $f(s)/s > A$. Then, from (17), it follows that for all $t \in (0, s)$

$$f(t)/t \ge \frac{n-1}{n} \frac{f(s)}{s} > \frac{n-1}{n} A$$

which implies $\lim_{t \to 0} f(t)/t = \infty$, and the proof is complete.

It is interesting to speculate on when, if ever, $\lambda = \infty$. We will give a partial answer to this question by presenting an example of a stationary point process for which $\lambda = \infty$. This example is particularly interesting for, even though λ is infinite, the probability of an infinite number of events occurring in a finite interval is zero for this process.

Consider a Poisson process with rate α, but suppose that α is itself a random variable having a distribution F. Denote this counting process by $\{N(t), t \geq 0\}$. (A more formal definition of $\{N(t), t \geq 0\}$ would be that given α, $\{N(t), t \geq 0\}$ is a Poisson process with rate α.)

It follows that $\{N(t), t \geq 0\}$ has stationary increments; but generally speaking, it will not have independent increments. (Why?) Now,

$$P\{N(t) \geq 1\} = \int_0^\infty P\{N(t) \geq 1 \mid \alpha\} \, dF(\alpha)$$

$$= \int_0^\infty (1 - e^{-\alpha t}) \, dF(\alpha)$$

Therefore,

$$\lim_{t \to 0} \frac{P\{N(t) \geq 1\}}{t} = \lim_{t \to 0} \int_0^\infty \frac{(1 - e^{-\alpha t})}{t} \, dF(\alpha)$$

$$= \int_0^\infty \alpha \, dF(\alpha) \tag{18}$$

where the interchange of limit and integrand is justified since

$$\frac{|1 - e^{-\alpha t}|}{t} \leq \alpha.$$

Hence, λ will be infinite when the mean value of α is infinite, i.e., when the integral (18) diverges. Note, however, that even in this case,

$$\sum_{n=0}^\infty P\{N(t) = n\} = \sum_{n=0}^\infty \int_0^\infty e^{-\alpha t} \frac{(\alpha t)^n}{n!} \, dF(\alpha)$$

Since all terms are nonnegative, we may interchange the sum and integral in the above to obtain

$$\sum_{n=0}^\infty P\{N(t) = n\} = \int_0^\infty \sum_{n=0}^\infty e^{-\alpha t} \frac{(\alpha t)^n}{n!} \, dF(\alpha)$$

$$= \int_0^\infty dF(\alpha)$$

$$= 1$$

Hence, even when $\lambda = \infty$, the probability of an infinite number of events in a finite interval is 0.

Now for any stationary point process $\{N(t), t \geq 0\}$, we have that

$$EN(t + s) = EN(t) + EN(s)$$

which implies (see 1.2) that

$$EN(t) = \mu t$$

for some $\mu \geq 0$.

What is the relationship between μ and λ? Since

$$\mu = \sum_{n=1}^{\infty} n P_n(t)/t$$

$$\geq \sum_{n=1}^{\infty} P_n(t)/t = \frac{1 - P_0(t)}{t}$$

it follows that $\mu \geq \lambda$. In order to determine when $\lambda = \mu$, we need to introduce the following concept. A stationary point process is said to be *regular* or *orderly* if

$$P\{N(t) \geq 2\} = o(t) \tag{19}$$

Thus, for instance, the Poisson process may be defined as a regular stationary point process having independent increments.

It should be noted that for a stationary point process, (19) implies that the probability that two or more events will occur simultaneously at any point is 0. To see this, divide the interval $[0, 1]$ into n equal parts. The probability of a simultaneous occurrence of events is less than the probability of two or more events in any of the intervals $(j/n, j + 1/n], j = 0, 1, \ldots, n - 1$; and hence this probability is bounded by $nP\{N(1/n) \geq 2\}$ which, by (19), goes to zero as $n \to \infty$. When $\mu < \infty$, the converse of this is also true. That is, any stationary point process for which μ is finite and for which there is zero probability of simultaneous events, is necessarily regular. The proof in this direction is more complicated and will not be given.

We will end this section by proving that $\lambda = \mu$ for a regular stationary point process. The following is known as *Korolyook's theorem*.

Theorem 2.5

Consider a regular stationary point process. Here, the mean number of events per unit time, μ, and the intensity λ, defined by (14), are identical. The case $\lambda = \mu = \infty$ is not excluded.

PROOF. Define the following notation:

A_k for the event $\{N(1) > k\}$
B_{nj} for the event $\{N(j + 1/n) - N(j/n) \geq 2\}$
$B_n = \bigcup_{j=0}^{n-1} B_{nj}$
C_{nkj} for the event $\{N(j + 1/n) - N(j/n) \geq 1, N(1) - N(j + 1/n) = k\}$

Let $\varepsilon > 0$ and a positive integer m be given. From the assumed regularity of the process, it follows that

$$P(B_{nj}) < \frac{\varepsilon}{n(m+1)}, \qquad j = 0, 1, \ldots, n-1$$

for all sufficiently large n. Hence,

$$P(B_n) \leq \sum_{j=0}^{n-1} P(B_{nj}) \leq \frac{\varepsilon}{m+1}$$

Therefore,

$$P(A_k) = P(A_k \bar{B}_n) + P(A_k B_n)$$

$$\leq P(A_k \bar{B}_n) + \frac{\varepsilon}{m+1} \qquad (20)$$

where $\bar{B}_n$ is the complement of B_n. However, a little thought reveals that $A_k \bar{B}_n = \bigcup_{j=0}^{n-1} C_{nkj} \bar{B}_n$, and hence,

$$P(A_k \bar{B}_n) \leq \sum_{j=0}^{n-1} P(C_{nkj})$$

which together with (20) implies that

$$\sum_{k=0}^{m} P(A_k) \leq \sum_{j=0}^{n-1} \sum_{k=0}^{m} P(C_{nkj}) + \varepsilon$$

$$= \sum_{j=0}^{n-1} P\{N(j+1/n) - N(j/n) \geq 1, N(1) - N(j+1/n) \leq m\} + \varepsilon$$

$$\leq \sum_{j=0}^{n-1} P\{N(j+1/n) - N(j/n) \geq 1\} + \varepsilon \qquad (21)$$

$$= nP\{N(1/n) \geq 1\} + \varepsilon$$

$$\leq \lambda + 2\varepsilon$$

for all n sufficiently large. Now since (21) is true for all m, it follows that

$$\sum_{k=0}^{\infty} P(A_k) \leq \lambda + 2\varepsilon$$

Hence,

$$\mu = EN(1) = \sum_{k=0}^{\infty} P\{N(1) > k\} = \sum_{k=0}^{\infty} P(A_k) \leq \lambda + 2\varepsilon$$

and the result is obtained as ε is arbitrary and it is already known that $\lambda \leq \mu$.

Problems

1. Show that Definition 2.1 implies Definition 2.2.
2. Show that Assumption (iv) of Definition 2.2 follows from Assumptions (ii) and (iii). [Hint. Derive a functional equation for $P\{N(t) = 0\}$.]
3. Suppose that each event has probability p of being registered, independently of other events. If the process that generates events is Poisson with rate λ, then show that the counting process of registered events is Poisson with rate λp.
4. Let $\{N(t), t \geq 0\}$ be a Poisson process with rate λ. Calculate $E[N(t) \cdot N(t + s)]$.
5. Suppose that $\{N_1(t), t \geq 0\}$ and $\{N_2(t), t \geq 0\}$ are independent Poisson processes with rates λ_1 and λ_2. Show that $\{N_1(t) + N_2(t), t \geq 0\}$ is Poisson with rate $\lambda_1 + \lambda_2$. Also, show that the probability that the first event of the combined process comes from $\{N_1(t), t \geq 0\}$ is $\lambda_1/\lambda_1 + \lambda_2$, *independently* of the time of the event.
6. For the processes in Problem 5, find the probability that the first process reaches n before the second reaches m.
7. Suppose $\{N(t), t \geq 0\}$ is a Poisson process with rate λ, and that Y is a positive random variable independent of $\{N(t), t \geq 0\}$. Determine $E[N(Y)]$, $\text{Var}[N(Y)]$.
8. For the infinite server Poisson queue of Section 4, let $Y(t)$ be the number of customers who have completed service by t. Show that $Y(t)$ and $X(t)$ are independent, and determine the distribution of $Y(t)$.
9. Let $S_1, S_2, \ldots$ be the arrival times for a Poisson process. Suppose $Y_1, Y_2, \ldots$ are independent and identically distributed random variables, independent of the Poisson process. Let $g(\ , \ )$ be a real-valued function of two variables, and define

$$X(t) = \sum_{i=1}^{N(t)} g(Y_i, S_i)$$

Determine the characteristic function of $X(t)$, and derive $E[X(t)]$ and $\text{Var}[X(t)]$.
10. Find the conditional distribution of $S_1, S_2, \ldots, S_n$ given that $S_n = t$.
11. Derive (11) and (12) by conditional mean and variance arguments.
12. Prove (13).
13. Let $\{N(t), t \geq 0\}$ be a nonhomogeneous Poisson process with mean value function $m(t)$. Show that, given $N(t) = n$, the unordered set of arrival times has the same distribution as n independent and identically distributed random variables having distribution function

$$F(x) = \begin{cases} \dfrac{m(x)}{m(t)} & x \leq t \\ 1 & x > t \end{cases}$$

14. Suppose that workmen incur accidents in accordance with a nonhomogeneous Poisson process with mean value function $m(t)$. Suppose further that each injured man is out of work for a random amount of time having distribution F. Let $X(t)$ be the number of workers who are out of work at time t. Find $\phi_{X(t)}(u)$ and by differentiating obtain $EX(t)$ and $\text{Var } X(t)$.
15. Consider a Poisson process with mean rate λ. Suppose that an event that occurs at time t is classified, independent of other events, into one of n categories

with probability $P_i(t), i = 1, 2, \ldots, n, \sum_1^n P_i(t) = 1$. Let $X_i(t)$ be the number of events during the time duration $(0, t)$ that belong to the ith category. Show that for each i, $\{X_i(t), t \geq 0\}$ is a nonhomogeneous Poisson process.

16. In Problem 15, show that the processes $\{X_i(t), t \geq 0\}$ are independent.

17. Suppose cars enter a one-way infinite highway at a Poisson rate λ. The ith car to enter chooses a velocity V_i and travels at this velocity. We assume that the V_i's are independent positive random variables having a common distribution F. Use Problem 15 to derive the distribution of the number of cars that are located in the interval (a, b) at time t. Assume that no time is lost when one car overtakes another car.

18. Consider a counting process with stationary, independent increments. Show that λ, as defined by (14), is infinite if and only if $P_0(t) = 0$ for all $t > 0$. Show that this latter implies that with probability 1, an infinite number of events will occur in every interval.

19. For λ and μ as defined in Section 2.5 give an example in which $\mu = \infty$, $\lambda < \infty$.

20. Show that if dispatching at $T/2$ maximizes the probability that the wait is less than y for every y, then it also minimizes the expected wait.

References

For further results on stationary point processes, see:

[1] CRAMÉR, H. and M. LEADBETTER, *Stationary and Related Stochastic Processes*, John Wiley and Sons, New York, (1966).

[2] KHINTCHINE, A., *Mathematical Methods in the Theory of Queueing*, Griffin Statistical Monographs, (1960).

3

RENEWAL THEORY

3.1. Introduction and Preliminaries

In the previous chapter we saw that the interarrival times for the Poisson process are independent and identically distributed exponential random variables. A natural generalization is to consider a counting process for which the interarrival times are independent and identically distributed with an arbitrary distribution. Such a counting process is called a *Renewal Process*.

Formally, let $\{X_n, n = 1, 2, \ldots\}$ be a sequence of nonnegative independent random variables with a common distribution F. To avoid trivialities, we suppose that $P\{X_n = 0\} < 1$. From the nonnegativity of X_n, it follows that EX_n exists, though it may be infinite, and we denote

$$\mu = \int_0^\infty x \, dF(x)$$

Let

$$S_0 = 0, \qquad S_n = \sum_1^n X_i, \qquad n \geq 1$$

and define

$$N(t) = \sup\{n : S_n \leq t\}$$

It follows from the strong law of large numbers that $S_n/n \to \mu$ with probability 1. Hence $S_n \leq t$ only finitely often, and so $N(t) < \infty$ with probability 1.

Definition 3.1

The process $\{N(t), t \geq 0\}$ is a *Renewal Process*.

We say that a renewal occurs at t if $S_n = t$ for some n. Since the interarrival times are independent and identically distributed, it follows that after each renewal the process starts over again.

31

Table 3.1 gives the intuitive interpretation of the random variables X_n, S_n and $N(t)$.

TABLE 3.1

Random Variable	Interpretation
X_n	time between the $(n-1)$st and nth renewal., i.e., the nth interarrival time
S_n	time of the nth renewal
$N(t)$	total number of renewals in $[0, t]$

We also have the important relationship that the number of renewals by time t is greater than or equal to n if and only if the nth renewal occurs by time t. Formally,

$$N(t) \geq n \Leftrightarrow S_n \leq t \tag{1}$$

From (1), we obtain

$$\begin{aligned}
P\{N(t) = n\} &= P\{N(t) \geq n\} - P\{N(t) \geq n + 1\} \\
&= P\{S_n \leq t\} - P\{S_{n+1} \leq t\} \\
&= F_n(t) - F_{n+1}(t)
\end{aligned} \tag{2}$$

Let

$$m(t) = EN(t)$$

$m(t)$ is called the *renewal function*, and much of renewal theory is concerned with determining its properties. The relationship between $m(t)$ and F is given by the following proposition.

Proposition 3.1

$$m(t) = \sum_{n=1}^{\infty} F_n(t) \tag{3}$$

PROOF.

$$N(t) = \sum_{n=1}^{\infty} A_n$$

where

$$A_n = \begin{cases} 1 & \text{if the } n\text{th renewal occurred in } [0, t] \\ 0 & \text{otherwise} \end{cases}$$

Hence,

$$EN(t) = E \sum_{1}^{\infty} A_n$$

$$= \sum_{1}^{\infty} EA_n$$

$$= \sum_{1}^{\infty} P\{A_n = 1\}$$

$$= \sum_{1}^{\infty} P\{S_n \le t\}$$

$$= \sum_{1}^{\infty} F_n(t)$$

where the interchange of expectation and summation is justified by the non-negativity of the A_n.

Proposition 3.2

$$m(t) < \infty \qquad \text{for all } t \ge 0$$

PROOF. Since $P\{X_n = 0\} < 1$, there exists an $\alpha > 0$ such that $P\{X_n \ge \alpha\} > 0$. Define a related renewal process $\{\overline{X}_n, n \ge 1\}$ by

$$\overline{X}_n = \begin{cases} 0 & \text{if } X_n < \alpha \\ \alpha & \text{if } X_n \ge \alpha \end{cases}$$

and let $\overline{N}(t) = \sup\{n : \overline{X}_1 + \cdots + \overline{X}_n \le t\}$. Then it is easy to see that for the related process, renewals can only take place at times $t = n\alpha, n = 0, 1, 2, \ldots,$ and also the number of renewals at each of these times are independent random variables with mean

$$\frac{1}{P\{X_n \ge \alpha\}}$$

Thus,

$$E\overline{N}(t) \le \frac{[t/\alpha] + 1}{P\{X_n \ge \alpha\}} < \infty$$

and the result follows since $\overline{X}_n \le X_n$ implies that $\overline{N}(t) \ge N(t)$.

REMARK. The above proof also shows that $E(N(t))^r < \infty$ for all $t \ge 0$, $r \ge 0$.

In Proposition 3.1, we show how $m(t)$ may be determined (at least in theory) from F. It also turns out that $m(t)$ uniquely determines F.

Proposition 3.3

There is a one-to-one correspondence between the interarrival distribution F and the renewal function $m(t)$.

PROOF. Taking Laplace transforms of both sides of the relationship $m(t) = \sum_{n=1}^{\infty} F_n(t)$ yields

$$\tilde{m}(s) = \sum_{n=1}^{\infty} \tilde{F}_n(s)$$

$$= \sum_{n=1}^{\infty} [\tilde{F}(s)]^n$$

$$= \frac{\tilde{F}(s)}{1 - \tilde{F}(s)} \tag{4}$$

or equivalently,

$$\tilde{F}(s) = \frac{\tilde{m}(s)}{1 + \tilde{m}(s)}$$

Hence $\tilde{F}$ is determined by $m(t)$; as the Laplace transform determines the distribution, it follows that F also is determined by $m(t)$.

3.2. Renewal Equation and Generalizations

An integral equation for $m(t)$ may be obtained by conditioning on the time of the first renewal. Upon doing so we obtain

$$m(t) = \int_0^{\infty} E[N(t) | X_1 = x] \, dF(x) \tag{5}$$

However,

$$E[N(t) | X_1 = x] = \begin{cases} 0 & x > t \\ 1 + m(t - x) & x \le t \end{cases} \tag{6}$$

for if the first renewal occurs at time x, $x \le t$, then from this point on the process starts over again, and thus the expected number of renewals in $[0, t]$ is just 1 plus the expected number to arrive in a time $t - x$ from the beginning of an equivalent renewal process.

Putting (6) in (5) yields

$$m(t) = \int_0^t [1 + m(t - x)] \, dF(x)$$

$$= F(t) + \int_0^t m(t - x) \, dF(x) \tag{7}$$

Equation (7) is known as the renewal equation, and can sometimes be solved for $m(t)$.

A generalization of the renewal equation is the following:

$$g(t) = h(t) + \int_0^t g(t - x) \, dF(x) \qquad (t \geq 0) \tag{8}$$

where h and F are known and g is an unknown function to be determined as a solution to the integral equation (8). The integral equation (8) is said to be a *renewal-type equation* and its solution is given by the following.

Proposition 3.4

If

$$g(t) = h(t) + \int_0^t g(t - x) \, dF(x) \qquad (t \geq 0)$$

then

$$g(t) = h(t) + \int_0^t h(t - x) \, dm(x)$$

where

$$m(x) = \sum_{n=1}^{\infty} F_n(x)$$

PROOF. Equation (8) states that

$$g = h + g * F$$

and taking Laplace transforms yields

$$\tilde{g}(s) = \tilde{h}(s) + \tilde{g}(s)\tilde{F}(s)$$

or

$$\tilde{g}(s) = \frac{\tilde{h}(s)}{1 - \tilde{F}(s)}$$

which is equivalent to

$$\tilde{g}(s) = \tilde{h}(s)\left[1 + \frac{\tilde{F}(s)}{1 - \tilde{F}(s)}\right] = \tilde{h}(s) + \tilde{h}(s)\frac{\tilde{F}(s)}{1 - \tilde{F}(s)}$$

Recalling that $\tilde{m}(s) = \dfrac{\tilde{F}(s)}{1 - \tilde{F}(s)}$ [Equation (4)], we obtain

$$\tilde{g}(s) = \tilde{h}(s) + \tilde{h}(s)\tilde{m}(s)$$

or

$$\tilde{g} = \overbrace{h + h * m}$$

Since the Laplace transform uniquely determines the function, this yields the desired result.

3.3. Limit Theorems

The expression, $1/\mu = 1/\int_0^\infty x\, dF(x)$ is often called the *rate of the process* (where $1/\infty = 0$). The theoretical justification for this nomenclature is given by the following theorem

Theorem 3.5

With probability 1,

$$\frac{N(t)}{t} \to \frac{1}{\mu} \qquad \text{as } t \to \infty.$$

PROOF.   By definition of $N(t)$, it follows (see Figure 3.1) that

$$S_{N(t)} \leq t \leq S_{N(t)+1}$$

Hence,

$$\frac{S_{N(t)}}{N(t)} \leq \frac{t}{N(t)} \leq \frac{S_{N(t)+1}}{N(t)} \tag{9}$$

Figure 3.1. *x* Marks a Renewal

Now the strong law of large numbers states that with probability 1, $S_n/n \to \mu$ as $n \to \infty$. Since $N(t) \to \infty$ with probability 1 as $t \to \infty$, we obtain

$$\frac{S_{N(t)}}{N(t)} \to \mu \quad \text{as } t \to \infty \quad \text{(with probability 1)} \tag{10}$$

By the same argument, we also see that with probability 1,

$$\frac{S_{N(t)+1}}{N(t)} = \frac{S_{N(t)+1}}{N(t)+1} \cdot \frac{N(t)+1}{N(t)} \to \mu \cdot 1 = \mu \quad \text{as } t \to \infty \tag{11}$$

The result then follows from (9), (10) and (11).

Thus the average number of renewals per unit time converges to $1/\mu$. How about the expected average number of renewals per unit time? Is it true that $m(t)/t \to 1/\mu$? In order to answer this (it is true), we must first digress to Wald's fundamental equation of sequential analysis.

3.4. Wald's Equation

Let $X_1, X_2, \ldots$ be a sequence of independent random variables.

Definition 3.2

An integer-valued positive random variable N is said to be a stopping time for the sequence $X_1, X_2, \ldots$ if the event $\{N = n\}$ is independent of $X_{n+1}, X_{n+2}, \ldots$ for all $n = 1, 2, \ldots$.

Intuitively, we observe the X_n's one at a time and N denotes the time at which we stop. If $N = n$, then we have stopped after observing $X_1, \ldots, X_n$ and before observing $X_{n+1}, X_{n+2}, \ldots$.

EXAMPLE 1. Let X_n, $n = 1, 2, \ldots$ be independent and such that

$$P\{X_n = 0\} = P\{X_n = 1\} = \tfrac{1}{2}, \quad n = 1, 2, \ldots$$

Let

$$N = \min\{n : X_1 + \cdots + X_n = 10\}$$

Then N is a stopping time.

EXAMPLE 2. Let X_n, $n = 1, 2, \ldots$ be independent and such that

$$P\{X_n = -1\} = P\{X_n = 1\} = \tfrac{1}{2}$$

Then

$$N = \min \{n : X_1 + \cdots + X_n = 1\}$$

is a stopping time. (It will be shown in the next chapter that N is finite with probability 1.)

Theorem 3.6 (Wald's Equation)

If $X_1, X_2, \ldots$ *are independent and identically distributed random variables having finite expectations, and if* N *is a stopping time for* $X_1, X_2, \ldots$, *such that* $EN < \infty$, *then*

$$E \sum_1^N X_i = EN \cdot EX$$

PROOF. Letting

$$Y_n = \begin{cases} 1 & \text{if } N \geq n \\ 0 & \text{if } N < n \end{cases}$$

we have that

$$\sum_{n=1}^N X_n = \sum_{n=1}^\infty X_n Y_n$$

thus,

$$E \sum_{n=1}^N X_n = E \sum_{n=1}^\infty X_n Y_n = \sum_{n=1}^\infty E(X_n Y_n) \tag{12}$$

However, $Y_n = 1$ if and only if we have not stopped after successively observing $X_1, \ldots, X_{n-1}$. Therefore, Y_n is determined by $X_1, \ldots, X_{n-1}$ and is thus independent of X_n. From (12), we thus obtain

$$E \sum_{n=1}^N X_n = \sum_{n=1}^\infty EX_n EY_n$$

$$= EX \sum_{n=1}^\infty EY_n$$

$$= EX \sum_{n=1}^\infty P\{N \geq n\}$$

$$= EX \cdot EN$$

REMARKS

(i) In Equation 12 we interchanged expectation and summation without an attempt at justification. To justify this interchange, replace X_i by $|X_i|$ throughout. In this case, the interchange is justified as all terms are nonnegative. However, this implies that the original interchange is allowable by Lebesgue's Dominated Convergence Theorem.

(ii) That $EN = \sum_{n=1}^{\infty} P\{N \geq n\}$ can be seen as follows:

$$\sum_{n=1}^{\infty} P\{N \geq n\} = \sum_{n=1}^{\infty} \sum_{k=n}^{\infty} P\{N = k\}$$

$$= \sum_{k=1}^{\infty} \sum_{n=1}^{k} P\{N = k\} = \sum_{k=1}^{\infty} kP\{N = k\} = EN$$

For Example 1, Wald's equation implies

$$E[X_1 + \cdots + X_N] = \tfrac{1}{2}EN$$

However, $X_1 + \cdots + X_N = 10$ by definition of N, and so $EN = 20$.

An application of the conclusion of Wald's equation to Example 2 would yield $E[X_1 + \cdots + X_N] = EN \cdot EX$. However, $X_1 + \cdots + X_N = 1$ and $EX = 0$, and so we would arrive at a contradiction. Thus Wald's equation is not applicable, which yields the conclusion that $EN = \infty$.

3.5. Back to Renewal Theory

In order to apply Wald's equation to renewal theory, we must first discover a stopping time. The obvious candidate is clearly $N(t)$. However,

$$N(t) = n \Leftrightarrow X_1 + \cdots + X_n \leq t \quad \text{and} \quad X_1 + \cdots + X_n + X_{n+1} > t \quad (13)$$

and thus the event $\{N(t) = n\}$ is not independent of X_{n+1}. From this analysis we must sadly conclude that

$$N(t) \text{ is } not \text{ a stopping time.}$$

However, after reflecting on Equation (13), it becomes clear that while $N(t)$ is not,

$$N(t) + 1 \text{ } is \text{ a stopping time.}$$

This is true since

$$N(t) + 1 = n \Leftrightarrow N(t) = n - 1$$

$$\Leftrightarrow X_1 + \cdots + X_{n-1} \leq t \quad \text{and} \quad X_1 + \cdots + X_n > t$$

Thus, the event $\{N(t) + 1 = n\}$ depends only on $X_1, \cdots, X_n$ and is therefore independent of $X_{n+1}, X_{n+2}, \ldots$; and the result follows from the definition of a stopping time.

Moreover, $E[N(t) + 1] = m(t) + 1 < \infty$ and thus from Wald's equation we obtain $E[X_1 + \cdots + X_{N(t)+1}] = EX \cdot E[N(t) + 1]$, or the equivalent

Corollary 3.7

If $\mu < \infty$, then

$$E[S_{N(t)+1}] = \mu(m(t) + 1) \tag{14}$$

We are now in position to prove

Theorem 3.8 (The Elementary Renewal Theorem)

$$\frac{m(t)}{t} \to \frac{1}{\mu} \qquad \text{as } t \to \infty$$

PROOF.   Suppose first that $\mu < \infty$. Now (see Figure 3.1)

$$S_{N(t)+1} \geq t$$

and thus by Corollary 3.7,

$$\mu(m(t) + 1) \geq t$$

implying that

$$\lim_{t \to \infty} \inf \frac{m(t)}{t} \geq \frac{1}{\mu} \tag{15}$$

To go the other way, we fix a constant M, and define a new renewal process $\{\bar{X}_n, n = 1, 2, \ldots\}$ by letting

$$\bar{X}_n = \begin{cases} X_n & \text{if } X_n \leq M \\ M & \text{if } X_n > M \end{cases} \qquad n = 1, 2, \ldots$$

Let $\bar{S}_n = \sum_1^n \bar{X}_i$, and $\bar{N}(t) = \sup\{n : \bar{S}_n \leq t\}$. Since the interarrival times for this truncated renewal process are bounded by M, we obtain

$$\bar{S}_{\bar{N}(t)+1} \leq t + M$$

Hence by Corollary 3.7,

$$(\bar{m}(t) + 1)\mu_M \leq t + M$$

where $\mu_M = E\overline{X}_n$. Thus,

$$\limsup_{t \to \infty} \frac{\overline{m}(t)}{t} \leq \frac{1}{\mu_M}$$

Now, since $\overline{S}_n \leq S_n$, it follows that $\overline{N}(t) \geq N(t)$ and $\overline{m}(t) \geq m(t)$, thus

$$\limsup_{t \to \infty} \frac{m(t)}{t} \leq \frac{1}{\mu_M} \qquad (16)$$

Letting $M \to \infty$ yields

$$\limsup_{t \to \infty} \frac{m(t)}{t} \leq \frac{1}{\mu} \qquad (17)$$

and the result follows from (15) and (17).

When $\mu = \infty$, we again consider the truncated process; since $\mu_M \to \infty$ as $M \to \infty$, the result follows from (16).

A nonnegative random variable X is said to be lattice if there exists $d \geq 0$ such that $\sum_{n=0}^{\infty} P\{X = nd\} = 1$. That is, X is lattice if it only takes on integral multiples of some nonnegative number d. The largest d having this property is said to be the period of X. If X is lattice and F is the distribution function of X, then we say that F is lattice.

We shall state without proof the following theorem.

Theorem 3.9 (Blackwell's Theorem)

(i) *If F is not lattice, then*

$$m(t + a) - m(t) \to a/\mu \qquad \text{as } t \to \infty$$

   for all $a \geq 0$.

(ii) *If F is lattice period d, then* $\lim_{n \to \infty} P\{\text{renewal at } nd\} \to d/\mu$

A function $h(t)$, $t \geq 0$, is said to be directly Riemann integrable if

$$\sum_{n=1}^{\infty} |\overline{m}_n(a)| \qquad \text{and} \qquad \sum_{n=1}^{\infty} |\underline{m}_n(a)|$$

are finite, and

$$\lim_{a \to 0} a \sum_{n=1}^{\infty} \underline{m}_n(a) = \lim_{a \to 0} a \sum_{n=1}^{\infty} \overline{m}_n(a)$$

where $\overline{m}_n(a)[\underline{m}_n(a)]$ is the maximum (minimum) of h on the interval $[(n-1)a, na]$. A sufficient condition for h to be directly Riemann integrable is that

(i) $h(t) \geq 0$ for all $t \geq 0$
(ii) $h(t)$ is nonincreasing
(iii) $\int_0^\infty h(t)\, dt < \infty$.

The following theorem, known as the *key renewal theorem*, will also be stated without proof.

Theorem 3.10 (Key Renewal Theorem)

If F is not lattice, and if h(t) is directly Riemann integrable, then

$$\lim_{t \to \infty} \int_0^t h(t - x)\, dm(x) = \frac{1}{\mu} \int_0^\infty h(t)\, dt$$

The key renewal theorem is a very important and useful result, and one can only appreciate it by seeing it in use. The technique we shall employ is to attempt to derive a renewal type equation and then use Proposition 3.4 to put the equation in a form suitable for application of the key renewal theorem. In order to derive a renewal type equation, we condition on the time at which the process restarts itself.

However, before considering an example, it might be worthwhile to stop for a moment and reflect on Blackwell's theorem and on the key renewal theorem. How, for instance, would someone (D. Blackwell and W. Smith, in particular) come to think that such theorems might be true? In the case of Blackwell's theorem, it is clear what the theorem asserts, hence someone with the right intuition might indeed conjecture it to be true. The content of the key renewal theorem, however, is at first glance, not at all clear. In order to clarify it, let us start with Blackwell's theorem and reason as follows: By Blackwell's theorem, we have that

$$\lim_{t \to \infty} \frac{m(t + a) - m(t)}{a} = \frac{1}{\mu}$$

and hence,

$$\lim_{a \to 0} \lim_{t \to \infty} \frac{m(t + a) - m(t)}{a} = \frac{1}{\mu}$$

Now, assuming that we can justify interchanging the limits, and assuming that everything is " nice," we obtain

$$\lim_{t \to \infty} \frac{dm(t)}{dt} = \frac{1}{\mu}$$

The key renewal theorem is just a formalization of this idea.

EXAMPLE 3. *Alternating Renewal Process.* Consider a system which can be in one of two states, *on* or *off*. Initially, it is on and it remains on for a time X_1; it then goes off and remains off for a time Y_1; it then goes on for a time X_2, then off for a time Y_2, then on, etc.

Suppose that the X_n's are independent, having a common distribution F, and the Y_n's are independent, having a common distribution G; suppose, furthermore, that X_n's and Y_n's are independent of each other. Let $H = F * G$, and let

$$P(t) = P\{\text{system is on at } t\}$$

Proposition 3.11

If $E(X + Y) < \infty$, and H is nonlattice, then

$$\lim_{t \to \infty} P(t) = \frac{EX}{EX + EY}$$

PROOF. Conditioning on $X_1 + Y_1$ yields

$$P(t) = \int_0^\infty P\{\text{on at } t \mid X_1 + Y_1 = x\} \, dH(x)$$

However, since at time $X_1 + Y_1$ the process restarts itself, we have

$$P\{\text{on at } t \mid X_1 + Y_1 = x\} = \begin{cases} P(t - x) & x \leq t \\ P\{X_1 > t \mid X_1 + Y_1 = x\} & x > t \end{cases}$$

Therefore,

$$P(t) = \int_0^t P(t - x) \, dH(x) + \int_t^\infty P\{X_1 > t \mid X_1 + Y_1 = x\} \, dH(x)$$

However, since $Y_1 \geq 0$, we have

$$\int_t^\infty P\{X_1 > t \mid X_1 + Y_1 = x\} \, dH(x) = \int_0^\infty P\{X_1 > t \mid X_1 + Y_1 = x\} \, dH(x)$$

$$= P\{X_1 > t\}$$

and we thus obtain the following renewal type equation:

$$P(t) = 1 - F(t) + \int_0^t P(t - x) \, dH(x)$$

Applying Proposition 3.4 yields

$$P(t) = 1 - F(t) + \int_0^t (1 - F(t - x)) \, dm_H(x)$$

where $m_H(x) = \sum_{n=1}^\infty H_n(x)$; and by applying the key renewal theorem with $h(t) = 1 - F(t)$, we obtain

$$\lim_{t \to \infty} P(t) = \frac{\int_0^\infty [1 - F(t)] \, dt}{\int_0^\infty x \, dH(x)} = \frac{EX}{EX + EY}$$

If we let $Q(t) = P\{\text{off at } t\} = 1 - P(t)$, then

$$Q(t) \to \frac{EY}{EX + EY}$$

and we note that the fact that the system was initially on makes no difference in the limit.

3.6. Excess Life and Age Distribution

Let $Y(t)$ be the time from t until the next renewal, and let $Z(t)$ be the time from t since the last renewal. That is,

$$Y(t) = S_{N(t)+1} - t$$
$$Z(t) = t - S_{N(t)}$$

$Y(t)$ is called the *excess* or *residual life* at t, and $Z(t)$ is called the *age* at t. In this section, we will determine the distributions of $Y(t)$ and $Z(t)$. We first note the important relationship,

$$Y(t) > x \Leftrightarrow \quad \text{no renewals in } [t, t + x] \tag{18}$$

Theorem 3.12

$$P\{Y(t) \le x\} = F(t + x) - \int_0^t [1 - F(t + x - y)]\, dm(y)$$

Furthermore, if F is not lattice, then

$$\lim_{t \to \infty} P\{Y(t) \le x\} = \frac{\int_0^x [1 - F(y)]\, dy}{\mu}$$

PROOF. Let $P(t) = P\{Y(t) > x\}$. Then by conditioning on X_1, we obtain

$$P(t) = \int_0^\infty P\{Y(t) > x \mid X_1 = s\}\, dF(s)$$

However, by keeping (18) in mind and using the fact that the process restarts itself at X_1, we obtain

$$P\{Y(t) > x \mid X_1 = s\} = \begin{cases} P(t - s) & s \le t \\ 0 & t < s \le t + x \\ 1 & s > t + x \end{cases}$$

Thus,

$$P(t) = \int_0^t P(t - s)\, dF(s) + \int_{t+x}^{\infty} dF(s)$$

$$= 1 - F(t + x) + \int_0^t P(t - s)\, dF(s)$$

and by applying Proposition 3.4 to this renewal type equation, we obtain

$$P(t) = 1 - F(t + x) + \int_0^t [1 - F(t + x - y)]\, dm(y)$$

On application of the key renewal theorem with $h(t) = 1 - F(t + x)$, we arrive at

$$\lim_{t \to \infty} P(t) = \int_0^{\infty} [1 - F(t + x)]\, dt/\mu$$

$$= \int_x^{\infty} [1 - F(y)]\, dy/\mu$$

or

$$\lim_{t \to \infty} P\{Y(t) \leq x\} = 1 - \int_x^{\infty} [1 - F(y)]\, dy/\mu$$

$$= \int_0^x [1 - F(y)]\, dy/\mu$$

where we have used the fact that $\mu = \int_0^{\infty} [1 - F(y)]\, dy$

To obtain the distribution of the age $Z(t)$ is easy. We just note that

$$Z(t) > x \Leftrightarrow \text{no renewals in } [t - x, t]$$

$$\Leftrightarrow Y(t - x) > x$$

Therefore,

$$P\{Z(t) > x\} = P\{Y(t - x) > x\}$$

and from Theorem 3.12, we have

Corollary 3.13

$$P\{Z(t) \leq x\} = \begin{cases} F(t) - \int_0^{t-x} [1 - F(t - y)]\, dm(y) & x < t \\[2mm] 1 & x \geq t \end{cases}$$

If F is not lattice, then

$$\lim_{t \to \infty} P\{Z(t) \leq x\} = \int_0^x [1 - F(y)]\, dy/\mu$$

3.7. Delayed Renewal Processes

We often consider a counting process for which the first interarrival time has a different distribution from the remaining ones. For instance, we might start observing a renewal process at some time $t > 0$. If a renewal does not occur at t, then the distribution of the time that we must wait until the first observed renewal will not be the same as the remaining interarrival distributions.

Formally, let $\{X_n, n = 1, 2, \ldots\}$ be a sequence of independent nonnegative random variables with X_1 having distribution G, and X_n having distribution F, $n > 1$. Let $S_0 = 0$, $S_n = \sum_1^n X_i$ $n \geq 1$, and define

$$N_D(t) = \sup\{n : S_n \leq t\}$$

DEFINITION.   The stochastic process $\{N_D(t), t \geq 0\}$ is called a *general or a delayed renewal process*.

When $G = F$ we have, of course, an ordinary renewal process. As in the ordinary case, we have

$$P\{N_D(t) = n\} = P\{S_n \leq t\} - P\{S_{n+1} \leq t\}$$
$$= G * F_{n-1}(t) - G * F_n(t)$$

Let

$$m_D(t) = EN_D(t)$$

Then it is easy to show that

$$m_D(t) = \sum_{n=1}^{\infty} G * F_{n-1}(t) \tag{19}$$

and by taking transforms of (19), we obtain

$$\tilde{m}_D(s) = \frac{\tilde{G}(s)}{1 - \tilde{F}(s)} \tag{20}$$

Also, by conditioning on X_1, we obtain

$$m_D(t) = \int_0^\infty E[N_D(t) \mid X_1 = x] \, dG(x)$$

$$= \int_0^t [1 + m(t - x)] \, dG(x)$$

$$= G(t) + \int_0^t m(t - x) \, dG(x) \tag{21}$$

where $m(t)$ is the renewal function for the ordinary renewal process with distribution F.

By using the corresponding result for the ordinary renewal process, it is easy to prove similar limit theorems for the delayed process. We leave the proof of the following proposition for the reader.

Let $\mu = \int_0^\infty x \, dF(x)$.

Proposition 3.14

(i) With probability 1,

$$\frac{N_D(t)}{t} \to \frac{1}{\mu} \qquad \text{as } t \to \infty$$

(ii) $\dfrac{m_D(t)}{t} \to \dfrac{1}{\mu}$ as $t \to \infty$

(iii) If F is not lattice, then

$$m_D(t + a) - m_D(t) \to \frac{a}{\mu} \qquad \text{as } t \to \infty$$

(iv) If F and G are lattice with period d, then

$$P\{\text{renewal at } nd\} \to \frac{d}{\mu} \qquad \text{as } n \to \infty$$

When $\mu < \infty$, the distribution function

$$F_e(x) = \int_0^x [1 - F(y)] \, dy/\mu \qquad 0 \le x \tag{22}$$

is called the *equilibrium distribution* for F. The Laplace transform of F_e is given by

$$\tilde{F}_e(s) = \int_0^\infty e^{-sx} \, dF_e(x)$$

$$= \int_0^\infty e^{-sx} \frac{1 - F(x)}{\mu} \, dx \tag{23}$$

$$= \frac{1}{\mu s} - \int_0^\infty \frac{e^{-sx} \, dF(x)}{\mu s} \tag{24}$$

$$= \frac{1 - \tilde{F}(s)}{\mu s} \tag{25}$$

where we integrated by parts to get from (23) to (24).

The delayed renewal process with $G = F_e$ is called the *equilibrium renewal process* and is extremely important. For suppose that we start observing a renewal process at time t. Then the process we observe is a delayed renewal process whose initial distribution is the distribution of $Y(t)$. Thus, for t large, it follows from Theorem 3.12 that the observed process is the equilibrium renewal process. The stationarity of this process is proven in the next theorem.

Let $Y_D(t)$ denote the excess at t for a delayed renewal process.

Theorem 3.15

For the equilibrium renewal process,

(i) $m_D(t) = t/\mu$
(ii) $P\{Y_D(t) \le x\} = \int_0^x [1 - F(y)]\,dy/\mu$ *for all* $t \ge 0$
(iii) $\{N_D(t), t \ge 0\}$ *has stationary increments.*

PROOF. (i) From (20) and (25), we have that

$$\tilde{m}_D(s) = \frac{1 - \tilde{F}(s)}{\mu s[1 - \tilde{F}(s)]} = \frac{1}{\mu s}$$

However, simple calculus shows that $1/\mu s$ is the Laplace transform of the function $h(t) = t/\mu$, and thus by the uniqueness of transforms, we obtain

$$m_D(t) = t/\mu$$

(ii) For a delayed renewal process,

$$P\{Y_D(t) > x, N_D(t) = 0\} = 1 - G(t + x) \tag{26}$$

and for $n \ge 1$,

$$P\{Y_D(t) > x, N_D(t) = n\} = \int_0^\infty P\{Y_D(t) > x, N_D(t) = n \mid S_n = y\}\,dG * F_{n-1}(y)$$

$$= \int_0^t [1 - F(t + x - y)]\,dG * F_{n-1}(y) \tag{27}$$

Thus, from (26) and (27), we obtain

$$P\{Y_D(t) > x\} = \sum_{n=0}^\infty P\{Y_D(t) > x, N_D(t) = n\}$$

$$= 1 - G(t + x) + \sum_{n=1}^\infty \int_0^t (1 - F(t + x - y)\,dG * F_{n-1}(y)$$

$$= 1 - G(t + x) + \int_0^t [1 - F(t + x - y)]\,d\left[\sum_{n=1}^\infty G * F_{n-1}(y)\right]$$

$$= 1 - G(t + x) + \int_0^t [1 - F(t + x - y)]\,dm_D(y)$$

Now, using part (i) and letting $G = F_e$ yields

$$P\{Y_D(t) > x\} = 1 - \int_0^{t+x} \frac{[1 - F(y)] \, dy}{\mu} + \int_0^t [1 - F(t + x - y)] \frac{dy}{\mu}$$

$$= 1 - \int_0^x \frac{[1 - F(y)] \, dy}{\mu}$$

or

$$P\{Y_D(t) \le x\} = \int_0^x \frac{[1 - F(y)] \, dy}{\mu}$$

(iii) To prove (iii) we note that $N_D(t + s) - N_D(s)$ may be interpreted as the number of renewals in time t of a delayed renewal process, where the initial distribution is the distribution of $Y_D(s)$. The result then follows from (ii).

3.8. Counter Models

A source emits pulses in accordance with a Poisson Process with rate λ. A counter registers the pulses. However, after each registration, the counter becomes locked for a period of time. What happens when a pulse is emitted when the counter is locked determines the nature of the counter.

In a type I or *nonparalyzable* counter, pulses arriving when the counter is locked are neither registered nor do they affect the counter in any way. An example of a type I counter is the Geiger counter.

In a type II or *paralyzable* counter, pulses arriving when the counter is locked are not registered, but the counter remains locked during the additional locking times of these pulses. In other words, a pulse is registered if and only if at the time it arrives the locking times of all previously arriving particles have expired. An example of a type II counter is the scintillator.

In the parlance of queueing theory, a type I counter is equivalent to a Poisson arrival single server queue in which entering customers balk when the system is nonempty. A type II counter corresponds to an infinite server Poisson queue. In both cases, the counter being locked corresponds to the queueing system being nonempty.

Let Y_n, $n = 1, 2, \ldots$ denote the locking time of the nth arriving pulse. We suppose that the Y_n are independent random variables with a common distribution G, where $G(0) = 0$. The system will alternate between periods of time at which the counter is locked and period of times at which it is alive (i.e., not locked). Let A_n and D_n, $n = 1, 2, \ldots$, denote respectively the length of the nth alive time and the nth dead time period. Also let

$$Z_1 = A_1, \qquad Z_n = D_{n-1} + A_n, \qquad n > 1$$

Then Z_n denotes the time from the $(n - 1)$st registration to the nth registration (see Figure 3.2).

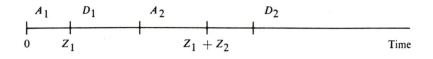

Figure 3.2.

It is often the case that the actual arrival rate λ is unknown and one desires to use the data (on registered pulses) to estimate λ. If we let $N_Z(t)$ denote the number of Z-renewals by t, we have that $\{N_Z(t), t \geq 0\}$ is a delayed renewal process. Thus by Proposition 3.14, we have for the observed rate $N_Z(t)/t$ that with probability 1,

$$\frac{N_Z(t)}{t} \to \frac{1}{EZ_2} = \frac{1}{EA + ED} = \frac{1}{1/\lambda + ED} \tag{28}$$

Thus from (28) it follows that an appropriate estimate for λ would be

$$\hat{\lambda} = \frac{N_Z(t)}{t - N_Z(t) \cdot ED} \tag{29}$$

In order to use (29) we must first calculate ED.

For a type I counter this is easy, as clearly $ED = EY$, and thus

$$\hat{\lambda} = \frac{N_Z(t)}{t - N_Z(t) \cdot EY}$$

For a type II counter we first note that due to the assumption of Poisson arrivals the system behaves as an alternating renewal process (see Figure 3.2). Thus letting $P_a(t)$ denote the probability that the counter is alive at t, we have by Proposition 3.11 that

$$\lim_{t \to \infty} P_a(t) = \frac{EA}{EA + ED} = \frac{1}{1 + \lambda ED}$$

or equivalently,

$$ED = \frac{1 - \lim_{t \to \infty} P_a(t)}{\lambda \lim_{t \to \infty} P_a(t)} \tag{30}$$

However,

$$P_a(t) = \sum_{n=0}^{\infty} P\{\text{Counter alive at } t | n \text{ pulses emitted in } (0, t)\} e^{-\lambda t} (\lambda t)^n / n! \tag{31}$$

Now, conditional on the event that there were n pulses emitted in $(0, t)$, the (unordered) set of times at which these pulses were emitted are distributed as n

independent uniform $(0, t)$ random variables. Hence the conditional probability that the locking times of all of these pulses have expired by t is given by

$$\left(\int_0^t G(t-x)\frac{dx}{t}\right)^n = \left(\int_0^t \frac{G(x)\,dx}{t}\right)^n$$

Thus, from (30) and (31), we obtain

$$P_a(t) = \sum_{n=0}^{\infty} e^{-\lambda t}\frac{(\lambda t)^n}{n!}\left(\int_0^t \frac{G(x)\,dx}{t}\right)^n$$

$$= \exp\left\{-\lambda \int_0^t [1 - G(x)]\,dx\right\}$$

and letting $t \to \infty$ yields

$$\lim_{t \to \infty} P_a(t) = \exp\left\{-\lambda \int_0^\infty [1 - G(x)]\,dx\right\} = e^{-\lambda EY} \qquad (32)$$

Thus using (32) and (30), we arrive at

$$ED = \frac{1 - e^{-\lambda EY}}{\lambda e^{-\lambda EY}} \qquad (33)$$

which unfortunately involves the unknown λ. However, by substituting $\hat{\lambda}$ for λ in (33), and using (29), we have that our estimate $\hat{\lambda}$ should approximately satisfy

$$\hat{\lambda} = \frac{N_Z(t)}{t - N_Z(t)\left(\dfrac{1 - e^{-\hat{\lambda} EY}}{\hat{\lambda} e^{-\hat{\lambda} EY}}\right)}$$

After simplifying, we obtain

$$\hat{\lambda} e^{-\hat{\lambda} EY} = \frac{N_Z(t)}{t} \qquad (34)$$

and (34) may be numerically solved for $\hat{\lambda}$.

3.9. Renewal Reward Process

Consider a renewal process with interarrival times $X_1, X_2, \ldots$. Suppose further that a reward Y_n is earned at the time of the nth renewal. Y_n may (and usually will) depend on X_n (the length of the renewal interval), but we suppose that the pairs (X_n, Y_n), $n = 1, 2, \ldots$ are independent and identically distributed. If we let

$$Y(t) = \sum_{n=1}^{N(t)} Y_n$$

then $Y(t)$ denotes the total reward earned by time t. The limiting value of the average return is given by the following theorem.

Theorem 3.16

If $E|Y_n|$ and EX_n are finite, then

(i) *with probability* 1,

$$Y(t)/t \to EY/EX \qquad \text{as } t \to \infty$$

(ii) $EY(t)/t \to EY/EX \qquad$ *as $t \to \infty$.*

PROOF. $Y(t)/t = \sum_{n=1}^{N(t)} Y_n/N(t) \cdot N(t)/t$, and (i) follows, since $\sum_{n=1}^{N(t)} Y_n/N(t) \to EY$ by the strong law of large numbers, and $N(t)/t \to 1/EX$.

To prove (ii) we first note that $N(t) + 1$ is a stopping time for $Y_1, Y_2, \ldots$, since the independence of $N(t) + 1$ and $\{X_{n+1}, X_{n+2}, \ldots\}$ implies the independence of $N(t) + 1$ and $\{Y_{n+1}, Y_{n+2}, \ldots\}$. Thus, by Wald's equation, we have

$$E \sum_{n=1}^{N(t)} Y_n = E \sum_{n=1}^{N(t)+1} Y_n - E[Y_{N(t)+1}]$$

$$= [m(t) + 1]EY - E[Y_{N(t)+1}]$$

$$\frac{EY(t)}{t} = \frac{m(t) + 1}{t} EY - \frac{E[Y_{N(t)+1}]}{t}$$

and the result will follow from the elementary renewal theorem if we can show that $1/tE[Y_{N(t)+1}] \to 0$ as $t \to \infty$. So, towards this end, let $g(t) = E[Y_{N(t)+1}]$. Conditioning on X_1 yields

$$g(t) = \int_0^\infty E[Y_{N(t)+1} \,|\, X_1 = x] \, dF(x)$$

However, it is easy to see that

$$E[Y_{N(t)+1} | X_1 = x] = \begin{cases} g(t - x) & x \le t \\ E[Y_1 | X_1 = x] & x > t \end{cases}$$

and thus we obtain the renewal type equation,

$$g(t) = \int_0^t g(t - x) \, dF(x) + h(t) \tag{35}$$

where $h(t) = \int_t^\infty E[Y_1|X_1 = x] \, dF(x)$. Also, note that since

$$E|Y_1| = \int_0^\infty E[|Y_1| \, | \, X_1 = x] \, dF(x) < \infty,$$

it follows that

$$h(t) \to 0 \quad \text{as } t \to \infty \quad \text{and } h(t) \le E|Y_1| \quad \text{for all } t \tag{36}$$

The renewal type equation (35) yields, by Proposition 3.4, the solution

$$g(t) = h(t) + \int_0^t h(t - x) \, dm(x)$$

By (36), we can choose T so that $|h(t)| < \varepsilon$ whenever $t \ge T$. Hence,

$$|g(t)|/t \le |h(t)|/t + \int_0^{t-T} \frac{|h(t - x)| \, dm(x)}{t} + \int_{t-T}^t \frac{|h(t - x)| \, dm(x)}{t}$$

$$\le \varepsilon/t + \varepsilon m(t - T)/t + E|Y_1| \frac{m(t) - m(t - T)}{t}$$

$$\to \varepsilon/EX \quad \text{as } t \to \infty$$

by the elementary renewal theorem. Since ε is arbitrary, it follows that $g(t)/t \to 0$, and the result follows.

REMARKS. If we say that a cycle is completed every time a renewal occurs, then the theorem states that the (expected) long-run average return is just the expected return earned during a cycle, divided by the expected time of a cycle.

In the proof of the theorem it is tempting to say that $E[Y_{N(t)+1}] = EY_1$, and thus $1/tE[Y_{N(t)+1}]$ trivially converges to zero. However, $Y_{N(t)+1}$ is related to $X_{N(t)+1}$, and $X_{N(t)+1}$ is the length of the renewal interval containing the point t. Since larger renewal intervals have a greater chance of containing t, it (heuristically) follows that $X_{N(t)+1}$ tends to be larger than an ordinary renewal interval (see problem 1), and thus the distribution of $Y_{N(t)+1}$ is not that of Y_1.

Also, up to now we have assumed that the reward is earned all at once at the end of the renewal cycle. However, this is not essential and Theorem 3.16 remains true if the reward is earned gradually during the renewal cycle. To see this, let $Y(t)$ denote the reward earned by t, and suppose first that all returns are nonnegative. Then

$$\frac{\sum_{n=1}^{N(t)} Y_n}{t} \le \frac{Y(t)}{t} \le \frac{\sum_{n=1}^{N(t)} Y_n}{t} + \frac{Y_{N(t)+1}}{t}$$

and (ii) of Theorem 3.16 follows, since

$$\frac{E[Y_{N(t)+1}]}{t} \to 0.$$

Part (i) of Theorem 3.16 follows by noting that both $\sum_{n=1}^{N(t)} Y_n/t$ and $\sum_{n=1}^{N(t)+1} Y_n/t$ converge to EY/EX by the argument given in the proof. A similar argument holds when the returns are nonpositive, and the general case follows by breaking up the returns into their positive and negative parts and applying the above argument separately to each.

EXAMPLE. Suppose that customers arrive at a train station in accordance with a renewal process with rate $1/\mu$. Whenever there are N customers in the station, a train leaves. Also, suppose that when there are n customers in the train station, the station incurs a holding cost at the rate of nc dollars per unit of time. What is the long-run average cost incurred by the train?

If we let X_i denote the time of departure of the ith train, and let Y_i denote the cost incurred between X_{i-1} and X_i, then it is clear that $\{(X_i, Y_i), i = 1, 2, \ldots\}$ is a renewal reward process. Hence, by Theorem 3.16, the average cost is EY_1/EX_1. Now

$$EY_1 = cE[\tau_1 + 2\tau_2 + \cdots + (N-1)\tau_{N-1}]$$

where τ_i is the time between the ith and $(i+1)$st customer arrival. Thus,

$$EY_1 = \frac{cN(N-1)\mu}{2}$$

implying, since $EX_1 = N\mu$, that the long-run average cost is

$$EY_1/EX_1 = \frac{c(N-1)}{2}$$

3.10 Nonterminating versus Terminating Renewal Processes

Up to this point, we have tacitly assumed that the interarrival distribution is an honest distribution, that is, that $F(\infty) = 1$. However, our definition 3.1 remains perfectly valid when $F(\infty) < 1$. Let

$$N(\infty) = \lim_{t \to \infty} N(t)$$

Lemma 3.17

If $F(\infty) = 1$, then $N(\infty) = \infty$ with probability 1.

PROOF. $N(\infty)$ is finite if and only if X_n is infinite for some n. Thus,

$$P\{N(\infty) < \infty\} \leq \sum_{n=1}^{\infty} P\{X_n = \infty\} = 0.$$

Proposition 3.18

$F(\infty) = 1 \Leftrightarrow EN(\infty) = \infty \Leftrightarrow N(\infty) = \infty$ with probability 1.

PROOF. If $F(\infty) = 1$, then by Lemma 3.17, $N(\infty) = \infty$ and thus $EN(\infty) = \infty$. If $F(\infty) < 1$ then after each renewal there is a positive probability $1 - F(\infty)$ that another renewal will not occur. Thus, $N(\infty)$ is a geometric random variable with finite mean $F(\infty)/1 - F(\infty)$, and so the result follows.

A renewal process with $F(\infty) = 1$ is called *recurrent or nonterminating*; one with $F(\infty) < 1$ is called *transient or terminating*.

REMARK. Proposition 3.18 is important, because often in applied probability models one does not start out with a renewal process (and thus with F), but rather one finds a renewal process *embedded* in some more general stochastic process. As a result, it is not always clear whether or not the interarrival distribution for this emdedded renewal process is honest. This will become clear in the next two chapters.

3.11. Age Dependent Branching Processes

Suppose that an organism at the end of its lifetime produces a random number of offspring in accordance with the probability distribution $\{P_j, j = 0, 1, 2, \ldots\}$. Assume further that all offspring act independently of each other and produce their own offspring in accordance with the same probability distribution $\{P_j\}$. Finally, let us assume that the lifetimes of the organisms are independent random variables with some common distribution F.

Let $X(t)$ denote the number of organisms alive at t. The stochastic process $\{X(t), t \geq 0\}$ is called an *age-dependent branching process*. We shall concern ourselves with determining the asymptotic form of $M(t) = EX(t)$, when $m = \sum_{j=0}^{\infty} jP_j > 1$.

Theorem 3.19

If $X_0 = 1$, $m > 1$ and F is not lattice, then

$$e^{-\alpha t}M(t) \to \frac{m-1}{m^2\alpha \int_0^\infty xe^{-\alpha x}\,dF(x)} \qquad \text{as } t \to \infty$$

where α is the unique positive number such that

$$\int_0^\infty e^{-\alpha x}\,dF(x) = 1/m$$

PROOF. By conditioning on T_1, the lifetime of the initial organism, we obtain

$$M(t) = \int_0^\infty E[X(t) \mid T_1 = s] \, dF(s) \tag{37}$$

However,

$$E[X(t)|T_1 = s] = \begin{cases} 1 & \text{if } s > t \\ m \cdot M(t - s) & \text{if } s \leq t \end{cases} \tag{38}$$

To see why (38) is true, suppose that $T_1 = s$, $s \leq t$ and suppose further that the organism has j offspring. Then the number of organisms alive at t may be written as $Y_1 + \cdots + Y_j$, where Y_i is the number of descendants (including himself) of the ith offspring that are alive at t. Clearly, $Y_1, \ldots, Y_j$ are independent with the same distribution as $X(t - s)$. Thus, $E(Y_1 + \cdots + Y_j) = jM(t - s)$; and (38) follows by taking the expectation (with respect to j) of $jM(t - s)$.

Thus, from (37) and (38), we obtain

$$M(t) = 1 - F(t) + m \int_0^t M(t - s) \, dF(s) \tag{39}$$

Equation (39) is almost a renewal type equation, and to transform it into one, we let α denote the unique positive number such that $\tilde{F}(\alpha) = 1/m$. Now, define a new distribution F_α by letting

$$F_\alpha(s) = m \int_0^s e^{-\alpha y} \, dF(y) \tag{40}$$

By multiplying both sides of (39) by $e^{-\alpha t}$ and using the fact that (40) implies that $dF_\alpha(s) = me^{-\alpha s} \, dF(s)$, we obtain

$$e^{-\alpha t} M(t) = e^{-\alpha t}(1 - F(t)) + \int_0^t e^{-\alpha(t-s)} M(t - s) \, dF_\alpha(s) \tag{41}$$

Letting $g(t) = e^{-\alpha t} M(t)$, it follows from Proposition 3.4 that the solution of the renewal type Equation (41) is

$$g(t) = e^{-\alpha t}[1 - F(t)] + \int_0^t e^{-\alpha(t-s)}[1 - F(t - s)] \, dm_\alpha(s)$$

and by the key renewal theorem,

$$g(t) \to \frac{\int_0^\infty e^{-\alpha t}[1 - F(t)] \, dt}{\int_0^\infty x \, dF_\alpha(x)} \tag{42}$$

Now, from integration by parts, it follows that

$$\int_0^\infty e^{-\alpha t}[1 - F(t)]\,dt = -[1 - F(t)]\frac{e^{-\alpha t}}{\alpha}\bigg|_{t=0}^{t=\infty} - \int_0^\infty \frac{e^{-\alpha t}}{\alpha}\,dF(t) \qquad (43)$$

$$= \frac{m-1}{m\alpha}\,.$$

Also,

$$\int_0^\infty x\,dF_\alpha(x) = m\int_0^\infty xe^{-\alpha x}\,dF(x) \qquad (44)$$

Thus from (42), (43) and (44), we obtain

$$e^{-\alpha t}M(t) \to \frac{m-1}{m^2\alpha\int_0^\infty xe^{-\alpha x}\,dF(x)} \qquad \text{as } t \to \infty$$

Problems

1. Let $X_{N(t)+1}$ be the length of the renewal interval containing t. Show that

$$P\{X_{N(t)+1} > x\} \ge 1 - F(x).$$

Find $P\{X_{N(t)+1} > x\}$ for a Poisson process.

2. Show that

$$E\{[(N(t)]^2\} = m(t) + 2\int_0^t m(t-s)\,dm(s)$$

3. The following Tauberian theorem relates the limiting properties of a function to the limiting properties of its Laplace transform.

THEOREM. *Let* $0 < \rho < 1 \infty$. *Then*

$$\lim_{t \to \infty} \frac{u(t)\Gamma(\rho+1)}{t^\rho} = 1 \Leftrightarrow \lim_{s \to 0} \tilde{u}(s)s^\rho = 1$$

where $\Gamma(\rho) = \int_0^\infty e^{-x}x^{\rho-1}\,dx$. [Note $\Gamma(n) = (n-1)!$] Use the above theorem to prove the elementary renewal theorem.

4. Consider a renewal process with $EX_1^2 < \infty$. Show that

$$\lim_{t \to \infty} P\left\{\frac{N(t) - t/\mu}{\sqrt{t\sigma^2/\mu^3}} < x\right\} = \frac{1}{\sqrt{2\pi}}\int_{-\infty}^x e^{-y^2/2}\,dy,$$

where $\sigma^2 = E(X_1 - \mu)^2$.

(Hint: Use the fact that $P\{N(t) < n\} = P\{S_n > t\}$.)

5. Show that (i) of Blackwell's theorem follows from the key renewal theorem, and give an example showing why we need to assume that F is not lattice in Part (i) of Blackwell's theorem.

6. A process may be in one of three states, 1, 2 or 3. It is initially in state 1, where it remains for an amount of time $\sim F_1$. After leaving state 1, it enters state 2, where it remains for an amount of time $\sim F_2$. It then goes to 3, where it remains a time $\sim F_3$. From 3, it returns to 1 and starts over again. Find

$$\lim_{t \to \infty} P \{\text{process is in state } i \text{ at } t\}, \qquad i = 1, 2, 3,$$

[Assume F_i are nonlattice and $\mu_i = \int_0^\infty x \, dF_i(x) < \infty$]. Show how to solve the n-state problem, where $1 \to 2 \to 3 \cdots \to n - 1 \to n \to 1$.

7. Show that a renewal process with exponential interarrival times is a Poisson process.

8. Let $Y(t)$ and $Z(t)$ denote respectively the excess and age of a renewal process.

    (a) $P\{Y(t) > x \,|\, Z(t) = s\} = ?$
    (b) $P\{Y(t) > x \,|\, Z(t + x/2) = s\} = ?$
    (c) for a Poisson process,

$$P\{Y(t) > x \,|\, Z(t + x) > s\} = ?$$

    (d) $P\{Y(t) > x, Z(t) > y\} = ?$

9. If $\mu < \infty$, show that with probability 1,

$$\frac{Z(t)}{t} \to 0 \qquad \text{as } t \to \infty$$

and also that

$$E \frac{Z(t)}{t} \to 0 \qquad \text{as } t \to \infty$$

10. For a nonlattice renewal process, if t is selected at random in the interval $(0, T)$, show that

$$\lim_{T \to \infty} P \{Y(t) \le x\} = \int_0^x \frac{(1 - F(y))}{\mu} \, dy$$

11. Let $Y(t)$ be the excess at t for a nonlattice renewal process. Find $\lim_{t \to \infty} EY(t)$ and show that it is equal to the mean of the limiting distribution of $Y(t)$. Assume $EX_1^2 < \infty$.

12. Prove Proposition 3.14.

13. For a nonlattice renewal process, show that

$$m(t) - t/\mu \to \frac{EX_1^2}{2\mu^2} - 1 \qquad \text{as } t \to \infty$$

Hint: Prove that

$$m(t) - t/\mu = -F_e(t) + \int_0^t [1 - F_e(t - x)] \, dm(x)$$

and apply the key renewal theorem.

14. Arrivals at a counter constitute a renewal process with interarrival distribution F. After each registration, the counter is locked for a fixed time L, during which all arrivals are without effect. Determine the distribution of the time from the end of a locked period until the next arrival. What if $F(x) = 1 - e^{-\lambda x}$?

15. Consider a counter in which arrivals come at a Poisson rate λ. The nth arriving particle locks the counter for a time T_n, and annuls the after-effect (if any) of its predecessors. The T_n, $n = 1, 2, \ldots$, are independent with a common distribution G. Let Y be the duration of a locked interval, and let $V(t) = P\{Y > t\}$. Show that

$$V(t) = [1 - G(t)]e^{-\lambda t} + \int_0^t V(t - x)[1 - G(x)]\lambda e^{-\lambda x}\, dx$$

16. The lifetime of a car is a random variable with distribution F. Mr. Moroney has a policy that he buys a new car as soon as his old car has either broken down or has reached the age of A years. Let τ denote the first time that Mr. Moroney replaces his car before it breaks down. Show that

$$E\tau = A + \frac{\int_0^A x\, dF(x)}{1 - F(A)}$$

Letting $V(t) = P\{\tau < t\}$, show that $V(t)$ satisfies

$$V(t) = h(t) + \int_0^t V(t - x)\, dG(x)$$

and determine $G(\ )$ and $h(\ )$.

17. For the preceding problem, suppose that a new car costs C_1 dollars and also that an additional cost of C_2 dollars is incurred whenever Mr. Moroney's car breaks down. Determine Mr. Moroney's long-term average cost per time.

18. *An Inventory Model.* Consider a store that sells a certain commodity. The weekly demands for this commodity are independent and have a common distribution G. At the beginning of each week, the store must decide how many additional units to order; in doing so, the store employs the following (s, S) ordering policy: If its present supply is not greater than s, then it orders enough to bring the supply up to S, and if the supply is greater than s then it does not order. In other words, its supply is x, then it orders

$$\begin{cases} S - x & \text{if } x \leq s \\ 0 & \text{if } x > s \end{cases}$$

The order is assumed to be filled immediately. Suppose that each demand which cannot be immediately met is lost, and we incur a penalty cost P. Also, presume a cost of C for each unit ordered. Explain how to derive the long-run average cost incurred, and derive it when $G(n) = 1 - \alpha^n$, $n = 0, 1, 2, \ldots$.

19. For the renewal reward process of Section 3.9, state and prove the equivalent of Blackwell's theorem.

20. Prove Theorem 3.16 when (X_1, Y_1) has a different distribution from (X_n, Y_n), $n > 1$. Assume that EX_1 and EY_1 are both finite. This is a general or delayed renewal reward process.

21. For the example of Section 3.9, suppose that a train is summoned whenever there are N customers present, but that it takes K units of time for the train to arrive at the station. When it arrives, it picks up all waiting customers. If the arriving stream of customers is a Poisson process, find the long-run average cost.

22. Solve Problem 21 when K is a random variable.

23. Consider a discrete time stochastic process which is initially in state 0. Each time the process returns to 0, it begins anew. That is, returns to 0 constitute renewals. Let P_n be the probability that the process will be in state 0 at time n, and show that returns to state 0 constitute a terminating renewal process if and only if $\sum_{n=1}^{\infty} P_n < \infty$. Find the distribution of the total number of renewals.

24. Find $P\{$for some $t > 0,\ N(t) = t\}$, when $\{N(t),\ t \geq 0\}$ is a Poisson process with rate λ.

25. Consider an age-dependent branching with which $X_0 = 1$ and $F(x) = 1 - e^{-\lambda x}$. Let $H(t)$ denote the probability that no organism is alive at t. Show that

$$H(t) = P_0(1 - e^{-\lambda t}) + \sum_{j=1}^{\infty} P_j \int_0^t [H(t - x)]^j \lambda e^{-\lambda x}\, dx$$

26. For the process in Problem 25, let L denote the sum of the lifetimes of all the organisms that have ever existed. Assuming that $m = \sum_{j=0}^{\infty} jP_j < 1$, show that

$$EL = \frac{1}{\lambda(1 - m)}$$

A very useful and stimulating review paper in renewal theory is the paper by Smith [5]. For instance, our proof of the elementary renewal theorem is taken from there. The key renewal theorem is also due to Smith (for a nice proof see Feller [1]). Theorem 3.15 is due to Doob. For further results on counter models, the reader is referred to Pyke [4]. Theorem 3.16 is not new (though the proof apparently is); it may be found in Johns and Miller [2]. The ideas of Section 3.10 are due to Feller. The proof of theorem 3.19 is due to Smith [5].

References

[1] FELLER, W., *An Introduction to Probability Theory and its Applications*, Vol. II, Wiley, New York, (1966).

[2] JOHNS, M. and R. MILLER., "Average Renewal Loss Rate," *Annals of Mathematical Statistics*, **34**, pp. 396–401, (1963).

[3] PRABHU, N. U., *Stochastic Processes*, Macmillan, New York, (1965).

[4] PYKE, R., "On Renewal Processes Related to Type I and Type II Counter Models," *Annals of Mathematical Statistics*, **29**, pp. 737–754, (1958).

[5] SMITH, W., "Renewal Theory and Its Ramifications," *Journal of the Royal Statistical Society*, Series B, **20**, No. 2, pp. 243–302, (1958).

4

MARKOV CHAINS

4.1. Preliminaries and Examples

A stochastic process $\{X_n, n = 0, 1, 2, \ldots\}$, with a finite or countable state space, is said to be a Markov chain if for all states $i_0, i_i, \ldots, i_{n-1}, i, j$, and all $n \geq 0$,

$$P\{X_{n+1} = j | X_0 = i_0, X_1 = i_1, \ldots, X_{n-1} = i_{n-1}, X_n = i\}$$
$$= P\{X_{n+1} = j | X_n = i\} \tag{1}$$

If $P\{X_{n+1} = j | X_n = i\}$ is independent of n, then the Markov chain is said to possess stationary transition probabilities. We shall only consider Markov chains with this property, and we shall let

$$P_{ij} = P\{X_{n+1} = j | X_n = i\}$$

It is convenient to label the state space of the process by the nonnegative integers $\{0, 1, 2, \ldots\}$ and we shall do so unless the contrary is explicitly stated. Also, we shall say that the process is in state j at time n whenever $X_n = j$.

A stochastic process satisfying (1) is said to possess the Markovian property. This property is equivalent to the statement that the conditional probability of any future event ($X_{n+1} = j$), given any past event ($X_0 = i_0, \ldots, X_{n-1} = i_{n-1}$), and the present state ($X_n = i$), is independent of the past event.

Let us denote by P the matrix of one-step transition probabilities P_{ij},

$$P = \left\| \begin{matrix} P_{00} & P_{01} & P_{02} & \cdots \\ P_{10} & P_{11} & P_{12} & \cdots \\ \vdots & \vdots & \vdots & \\ P_{n0} & P_{n1} & P_{n2} & \cdots \\ \vdots & \vdots & \vdots & \end{matrix} \right\|$$

Because of the interpretation of the P_{ij}'s as probabilities, we have, of course, that

$$P_{ij} \geq 0, \, i, j = 0, 1, 2, \ldots; \quad \sum_{j=0}^{\infty} P_{ij} = 1, \, i = 0, 1, \ldots$$

It turns out that the process is completely specified once P and the initial probability distribution of X_0 are given. To see this, we note that

$$P\{X_0 = i_0, X_1 = i_1, \ldots, X_n = i_n\}$$

$$= P\{X_n = i_n \mid X_0 = i_0, \ldots, X_{n-1} = i_{n-1}\}$$

$$P\{X_0 = i_0, \ldots, X_{n-1} = i_{n-1}\}$$

$$= P_{i_{n-1} i_n} \cdot P\{X_0 = i_0, \ldots, X_{n-1} = i_{n-1}\}$$

$$= P_{i_{n-1} i_n} \cdot P_{i_{n-2} i_{n-1}} \cdot P\{X_0 = i_0, \ldots, X_{n-2} = i_{n-2}\}$$

$$\vdots$$

$$= P_{i_{n-1} i_n} \cdot P_{i_{n-2} i_{n-1}} \cdot \cdots \cdot P_{i_0, i_1} \cdot P\{X_0 = i_0\}$$

Let P_{ij}^n represent the probability that the process goes from state i to state j in n steps; or formally,

$$P_{ij}^n = P\{X_{n+m} = j \mid X_m = i\}$$

Proposition 4.1

For any $r \leq n$,

$$P_{ij}^n = \sum_{k=0}^{\infty} P_{ik}^r P_{kj}^{n-r} \tag{2}$$

PROOF.   Since the process must be somewhere at time r, we have

$$P\{X_n = j \mid X_0 = i\} = \sum_{k=0}^{\infty} P\{X_n = j, X_r = k \mid X_0 = i\}$$

$$= \sum_{k=0}^{\infty} P\{X_n = j \mid X_r = k, X_0 = i\} \cdot P\{X_r = k \mid X_0 = i\}$$

$$= \sum_{k=0}^{\infty} P_{kj}^{n-r} P_{ik}^r$$

Equation (2) is known as the *Chapman-Kolmogorov equation*. If we let $P^{(n)}$ denote the matrix of n-step transition probabilities P_{ij}^n, then Equation (2) asserts that

$$P^{(n)} = P^{(r)} \cdot P^{(n-r)}$$

where the dot represents matrix multiplication. Hence,

$$P^{(n)} = P \cdot P^{(n-1)} = P \cdot P \cdot P^{n-2} = \cdots = P^n$$

and thus $P^{(n)}$ may be calculated by multiplying the matrix P by itself n times.

EXAMPLE 1. *Sums of Independent, Identically Distributed, Random Variables.* Let ξ_i, $i = 1, 2, \ldots$ be independent and identically distributed with

$$P\{\xi_1 = j\} = a_j, \qquad j \geq 0$$

$$\sum_{j=0}^{\infty} a_j = 1$$

If we let $X_n = \sum_{i=1}^{n} \xi_i$, then $\{X_n, n = 1, 2, \ldots\}$ is a Markov chain for which

$$P_{ij} = \begin{cases} a_{j-i} & j \geq i \\ 0 & j < i \end{cases}$$

One possible interpretation of X_n is the following: If ξ_i represents the demand for a commodity during the ith period, then X_n would represent the total demand for the first n periods.

EXAMPLE 2. *An Inventory Model.* Consider a store that stocks a certain commodity, and suppose that the weekly demands for this commodity are independent and identically distributed with

$$P\{\text{weekly demand} = j\} = a_j, \qquad j \geq 0$$

Suppose that the store employs the following (s, S) ordering policy. If at the beginning of the week, its present supply is s or higher, then it does not order; while if its supply is less than s, then the store orders enough to bring the supply up to S. That is, if its supply is x, then it orders

$$\begin{cases} 0 & \text{if } x \geq s \\ S - x & \text{if } x < s \end{cases}$$

The order is assumed to be filled immediately.

Suppose that all demands which cannot be immediately filled are lost, and let X_n be the inventory supply at the end of the nth week. Then

$$X_{n+1} = \begin{cases} \max(X_n - \xi_{n+1}, 0) & \text{if } X_n \geq s \\ \max\{S - \xi_{n+1}, 0\} & \text{if } X_n < s \end{cases}$$

where ξ_n is the demand for the nth week. Hence, $\{X_n, n = 0, 1, 2, \ldots\}$ is a Markov chain having

$$
P_{ij} = \begin{cases}
\displaystyle\sum_{k=i}^{\infty} a_k & \text{if } j = 0, & i \geq s \\[2ex]
a_{i-j} & \text{if } 0 < j \leq i, & i \geq s \\[2ex]
\displaystyle\sum_{k=S}^{\infty} a_k & \text{if } j = 0, & i < s \\[2ex]
a_{S-j} & \text{if } 0 < j \leq S, & i < s \\[2ex]
0 & \text{otherwise}
\end{cases}
$$

EXAMPLE 3. *The M/G/1 Queue.* Suppose that customers arrive at a service center in accordance with a Poisson process with mean rate λ. The customers are served one at a time by a single server, and if this server is busy when a customer arrives, then the customer waits on line. The service times for customers are assumed to be independent random variables with a common distribution G.

Let X_n denote the number of customers left in the system after the nth completion of service. Also, let Y_n denote the number of customers arriving during the service period of the $(n + 1)$st customer. Then it is easy to see that

$$
X_{n+1} = \begin{cases}
X_n - 1 + Y_n & \text{if } X_n > 0 \\
Y_n & \text{if } X_n = 0
\end{cases}
\tag{3}
$$

Also because of the lack of memory of the Poisson process, it follows that $Y_n, n = 1, 2, \ldots$ are independent, identically distributed random variables, having

$$
P\{Y_n = j\} = \int_0^{\infty} e^{-\lambda x} \frac{(\lambda x)^j}{j!} \, dG(x), \qquad j = 0, 1, 2, \ldots
\tag{4}
$$

From (3) and (4), it follows that $\{X_n, n = 1, 2, \ldots\}$ is a Markov chain with transition probabilities given by

$$
P_{0j} = \int_0^{\infty} e^{-\lambda x} \frac{(\lambda x)^j}{j!} \, dG(x), \qquad j = 0, 1, 2, \ldots
$$

$$
P_{ij} = \int_0^{\infty} e^{-\lambda x} \frac{(\lambda x)^{j-i+1}}{(j-i+1)!} \, dGx, \qquad j \geq i - 1, i = 0, 1, 2, \ldots
$$

$$
P_{ij} = 0, \qquad j < i - 1
$$

EXAMPLE 4. *The G/M/1 Queue.* Suppose now that customers arrive in accordance with an arbitrary renewal process, but the service distribution is exponential with mean $1/\lambda$.

Let X_n denote the number of customers in the system after the nth customer arrives. Then, letting Y_n denote the number of customers served during the time between the nth and $(n + 1)$st arrival, we have

$$X_{n+1} = X_n + 1 - Y_n \qquad (5)$$

Now, due to the lack of memory of the exponential, it follows that the distribution of Y_n depends only on X_n and not on $X_{n-1}, \ldots, X_1, X_0$. Also, for $j < X_n$, the event $\{Y_n = j\}$ is equivalent to having j Poisson events occurring in an arbitrary independent service time interval. Similarly, for $j = X_n$, the event $\{Y_n = j\}$ is equivalent to j or more Poisson events occurring in an arbitrary independent service time interval. Thus, letting G denote the interarrival distribution of customers, we arrive at

$$P\{Y_n = j \mid X_n = i, X_{n-1}, \ldots, X_0\} = \begin{cases} \int_0^\infty e^{-\lambda x} \dfrac{(\lambda x)^j}{j!} \, dG(x) & j < i \\[2ex] \sum\limits_{k=i}^{\infty} \int_0^\infty e^{-\lambda x} \dfrac{(\lambda x)^k}{k!} \, dG(x) & j = i \\[2ex] 0 & j > i \end{cases}$$

$$(6)$$

Hence, from (5) and (6), it follows that $\{X_n, n = 0, 1, 2, \ldots\}$ is a Markov chain with transition probabilities given by

$$P_{i1} = \sum_{k=i}^{\infty} \int_0^\infty e^{-\lambda x} \frac{(\lambda x)^k}{k!} \, dG(x)$$

$$P_{ij} = \int_0^\infty e^{-\lambda x} \frac{(\lambda x)^{i+1-j}}{(i+1-j)!} \, dG(x) \qquad i + 1 \geq j > 1$$

$$P_{ij} = 0 \qquad\qquad\qquad j > i + 1 \quad \text{or } j = 0$$

The reader should note that in the previous two examples we were able, by looking at the process only at certain time points, and by choosing these time points so as to exploit the lack of memory of the exponential distribution, to discover an *embedded* Markov chain. This is often a fruitful approach for processes in which the exponential distribution is present.

4.2. Classification of States

State j is said to be accessible from state i if for some $n \geq 0$, $P_{ij}^n > 0$. Two states i and j accessible to each other are said to *communicate*, and we write $i \leftrightarrow j$.

Proposition 4.2

Communication is an equivalence relation. That is,

(i) $i \leftrightarrow i$
(ii) if $i \leftrightarrow j$, then $j \leftrightarrow i$
(iii) if $i \leftrightarrow j$ and $j \leftrightarrow k$, then $i \leftrightarrow k$.

PROOF. The first two parts follow trivially from the definition of communication. To prove (iii), suppose that $i \leftrightarrow j$ and $j \leftrightarrow k$; then there exists m, n such that $P_{ij}^m > 0$, $P_{jk}^n > 0$. Hence,

$$P_{ik}^{m+n} = \sum_{r=0}^{\infty} P_{ir}^m P_{rk}^n \geq P_{ij}^m P_{jk}^n > 0$$

Similarly, we may show there exists an s for which $P_{ki}^s > 0$.

Two states which communicate are said to be in the same *class*; and by Proposition 4.2, any two classes are either disjoint or identical. We say that the Markov chain is *irreducible* if there is only one class, that is, if all states communicate with each other.

State i is said to have period d if $P_{ii}^n = 0$ except when $n = d, 2d, 3d, \ldots$ and d is the greatest integer with this property. (If $P_{ii}^n = 0$ for all n, then define the period of i to be zero.) A state with period one is said to be *aperiodic*. Let $d(i)$ denote the period of i. We now show that periodicity is a class property.

Proposition 4.3

If $i \leftrightarrow j$, then $d(i) = d(j)$.

PROOF. Let m and n be such that $P_{ij}^m > 0$, $P_{ji}^n > 0$. Suppose now that $P_{ii}^s > 0$. Then $P_{jj}^{n+s+m} \geq P_{ji}^n P_{ii}^s P_{ij}^m > 0$. Also, $P_{ii}^s > 0$ implies that $P_{ii}^{2s} \geq P_{ii}^s P_{ii}^s > 0$, and the same reasoning shows that $P_{jj}^{n+2s+m} > 0$. Hence, $d(j)$ divides both $n + s + m$ and $n + 2s + m$, thus $n + 2s + m - (n + s + m) = s$, whenever $P_{ii}^s > 0$. Therefore, $d(j)$ divides $d(i)$. A similar argument yields that $d(i)$ divides $d(j)$, thus $d(i) = d(j)$.

For any states i and j, we define f_{ij}^n to be the probability that starting in i, the first transition into j occurs at time n. Or, formally,

$$f_{ij}^0 = 0$$
$$f_{ij}^n = P\{X_n = j, X_k \neq j, k = 1, 2, \ldots, n-1 \mid X_0 = i\}$$

Let T_j denote the time of the first transition into state j, and let it be infinite if such a transition never occurs. Then conditioning on T_j yields

$$P_{ij}^n = \sum_{k=0}^{\infty} P\{X_n = j \mid T_j = k, X_0 = i\} \cdot P\{T_j = k \mid X_0 = i\}$$

$$+ P\{X_n = j \mid T_j = \infty, X_0 = i\} \cdot P\{T_j = \infty \mid X_0 = i\}$$

$$= \sum_{k=0}^{n} P_{jj}^{n-k} f_{ij}^k$$

where we have used the Markovian property in asserting that

$$P\{X_n = j \mid T_j = k, X_0 = i\} = P_{jj}^{n-k}.$$

Let

$$f_{ij} = \sum_{n=1}^{\infty} f_{ij}^n$$

Then f_{ij} denotes the probability of ever making a transition into state j, given that the process starts in i. (Note that for $i \neq j$, f_{ij} is nonzero if and only if j is accessible from i.)

State j is said to be *recurrent* if $f_{jj} = 1$, and *transient* otherwise. Thus, j is recurrent if with probability one, a process starting at j will eventually return to j. However, by the Markovian property, it follows that the process restarts itself upon returning to j; in other words, transitions into j constitute renewals. Hence, state j being recurrent corresponds to the renewal distribution being honest; thus, by Proposition 3.18, state j is recurrent if and only if the expected number of transitions into j is infinite.

Proposition 4.4

State j is recurrent if and only if

$$\sum_{n=1}^{\infty} P_{jj}^n = \infty$$

PROOF. Let

$$A_n = \begin{cases} 1 & X_n = j \\ 0 & \text{if } X_n \neq j \end{cases}$$

Then $\sum_{n=1}^{\infty} A_n$ denotes the number of transitions into j. The result then follows, since $E \sum_{n=1}^{\infty} A_n = \sum_{n=1}^{\infty} E A_n = \sum_{n=1}^{\infty} P_{jj}^n$.

For $t \leq \infty$, let $N_j(t)$ denote the total number of transitions into state j, up to time t. Then it also follows from Proposition 3.18 that

$$P\{N_j(\infty) = \infty \,|\, X_0 = j\} = \begin{cases} 1 & \text{if } j \text{ is recurrent} \\ 0 & \text{if } j \text{ is transient} \end{cases}$$

Moreover, if $X_0 = j$ and j is transient, then $N_j(\infty)$ is geometric with mean $f_{jj}/(1 - f_{jj})$.

We may also show that recurrence, like periodicity, is a class property.

Corollary 4.5

If i is recurrent and $i \leftrightarrow j$, then j is recurrent.

PROOF.   Let m, n be such that $P_{ij}^m > 0$, $P_{ji}^n > 0$. Now, $P_{jj}^{m+n+s} \geq P_{ji}^n P_{ii}^s P_{ij}^m$, and thus $\sum_{s=1}^{\infty} P_{jj}^{m+n+s} \geq P_{ji}^n P_{ij}^m \sum_{s=1}^{\infty} P_{ii}^s = \infty$, and the result obtains.

Using Corollary 4.5, we may prove:

Corollary 4.6

If $i \leftrightarrow j$ and j is recurrent, then

$$f_{ij} = 1$$

PROOF.   Suppose $X_0 = i$, and let n be such that $P_{ij}^n > 0$. Say that we miss opportunity 1 if $X_n \neq j$. If we miss opportunity 1, then let T_1 denote the next time we enter i (T_1 is finite with probability 1 by Corollary 4.5). Say that we miss opportunity 2 if $X_{T_1+n} \neq j$. If opportunity 2 is missed, let T_2 denote the next time we enter i and say that we miss opportunity 3 if $X_{T_2+n} \neq j$, etc. It is easy to see that the number of missed opportunities before the first success is random with finite mean $(1 - P_{ij}^n)/P_{ij}^n$, and is thus finite with probability 1. The result follows, since i being recurrent implies that the number of potential opportunities is infinite.

Thus, if $X_0 = j$ and j is recurrent, then returns to j constitute a recurrent renewal process with interarrival distribution $\{f_{jj}^n, n = 1, 2, \ldots\}$. If $X_0 = i$, $i \leftrightarrow j$, and j is recurrent, then transitions into j constitute a recurrent delayed renewal process with initial renewal distribution $\{f_{ij}^n, n = 1, 2, \ldots\}$.

EXAMPLE 5.   *Unrestricted Random Walk.*   Let the state space be the integers, and suppose that $P_{n,n+1} = p = 1 - P_{n,n-1}$ ($n = 0, \pm 1, \pm 2 \ldots$). When $0 < p < 1$, it is clear that the chain is irreducible, and thus all states are recurrent if state 0 is recurrent. Now,

$$P_{00}^{2n+1} = 0, \qquad n = 0, 1, 2, \ldots$$

and

$$P_{00}^{2n} = \binom{2n}{n} p^n (1-p)^n = \frac{(2n)!}{n!\,n!}\, p^n (1-p)^n$$

By using Stirling's approximation,

$$n! \sim n^{n+1/2} e^{-n} \sqrt{2\pi}$$

we obtain

$$P_{00}^{2n} \sim \frac{(4p(1-p))^n}{\sqrt{\pi n}}$$

However, $p(1-p) = \frac{1}{4}$ with equality holding if and only if $p = \frac{1}{2}$. Hence, $\sum_{n=0}^{\infty} P_{00}^n = \infty$ if and only if $p = \frac{1}{2}$. Thus, the chain is recurrent when $p = \frac{1}{2}$ and transient otherwise. (If $p \neq \frac{1}{2}$, then it is clear by the strong law of large numbers that the chain could not be recurrent.)

One may also look at the symmetric random walk in more than one dimension. For instance, in the two dimensional random walk, the probability of a transition one unit to the right, one unit to the left, up, or down would all equal $\frac{1}{4}$. Similarly in three dimensions the probability of a transition to any one of the 6 adjacent points would all be $\frac{1}{6}$. By using the same method as in the previous example, one may prove the interesting fact that the two dimensional symmetric random walk is recurrent but that all higher dimensional symmetric random walks are transient.

4.3. Limit Theorems

It is easily shown that if state j is transient, then

$$\sum_{n=1}^{\infty} P_{ij}^n < \infty \qquad \text{for all } i$$

meaning that given $X_0 = i$, the expected number of transitions into state j is finite. As a consequence, it follows that for j transient, $P_{ij}^n \to 0$ as $n \to \infty$.

Let

$$\mu_{jj} = \begin{cases} \infty & \text{if } j \text{ is transient} \\ \sum_{n=1}^{\infty} n f_{jj}^n & \text{if } j \text{ is recurrent} \end{cases}$$

Then, by proposition 3.14 and the above, we have

Theorem 4.7

If $i \leftrightarrow j$, then

(i) $P\left\{\lim\limits_{t \to \infty} N_j(t)/t = 1/\mu_{jj} \mid X_0 = i\right\} = 1$

(ii) $\lim\limits_{n \to \infty} \sum\limits_{k=1}^{n} P_{ij}^k / n = 1/\mu_{jj}$

(iii) *j is aperiodic implies*

$$\lim_{n \to \infty} P_{ij}^n = 1/\mu_{jj}$$

(iv) *j has period d implies*

$$\lim_{n \to \infty} P_{jj}^{nd} = d/\mu_{jj}$$

Suppose that state j is recurrent. Then state j is said to be *positive recurrent* if $\mu_{jj} < \infty$, and *null recurrent* if $\mu_{jj} = \infty$. Letting $\pi_j = \lim_{n \to \infty} P_{jj}^{nd(j)}$, it follows that j is positive recurrent if $\pi_j > 0$, and null recurrent if $\pi_j = 0$. A positive recurrent, aperiodic state is called *ergodic*. By our usual technique, we may prove

Proposition 4.8

Positive (null) recurrence is a class property.

Definition 4.1

A probability distribution $\{P_j, j = 0, 1, 2, \ldots\}$ is called *stationary* if

$$P_j = \sum_{i=0}^{\infty} P_i P_{ij} \tag{7}$$

If the initial probability distribution $\{P_j, j = 0, 1, 2, \ldots\}$ is stationary, then

$$P\{X_1 = j\} = \sum_{i=0}^{\infty} P\{X_1 = j \mid X_0 = i\} P\{X_0 = i\}$$

$$= \sum_{i=0}^{\infty} P_i P_{ij} = P_j$$

and by induction,

$$P\{X_n = j\} = \sum_{i=0}^{\infty} P\{X_n = j \mid X_{n-1} = i\} P\{X_{n-1} = i\}$$

$$= \sum_{i=0}^{\infty} P_{ij} P_i = P_j \tag{8}$$

Hence, when the initial probability distribution is stationary, then X_n will have the same distribution for all n; that is, $\{X_n, n = 0, 1, \ldots\}$ will be a stationary process.

Theorem 4.9

An irreducible aperiodic Markov chain belongs to one of the following two classes:

   (a) *Either the states are all transient or all null recurrent; in this case, $P_{ij}^n \to 0$ as $n \to \infty$ for all i, j and there exists no stationary distribution.*

   (b) *Or else, all states are positive recurrent, that is,*

$$\pi_j = \lim_{n \to \infty} P_{ij}^n > 0$$

In this case, $\{\pi_j, j = 0, 1, 2, \ldots\}$ is a stationary distribution and there exists no other stationary distribution.

PROOF. To prove (b), we first note that

$$\sum_{j=0}^{\infty} \pi_j \leq 1$$

This is true as $\sum_{j=0}^{M} P_{ij}^n \leq \sum_{j=0}^{\infty} P_{ij}^n = 1$; thus letting $n \to \infty$ yields $\sum_{j=0}^{M} \pi_j \leq 1$ for all M. Similarly, $P_{ij}^{n+1} = \sum_{k=0}^{\infty} P_{ik}^n P_{kj} \geq \sum_{k=0}^{M} P_{ik}^n P_{kj}$, and letting $n \to \infty$ yields $\pi_j \geq \sum_{k=0}^{M} \pi_k P_{kj}$ for all M; thus,

$$\pi_j \geq \sum_{k=0}^{\infty} \pi_k P_{kj}; \qquad j = 0, 1, 2, \ldots \tag{9}$$

Suppose (9) is a strict inequality for some j. Then adding these inequalities yields

$$\sum_{j=0}^{\infty} \pi_j > \sum_{j=0}^{\infty} \sum_{k=0}^{\infty} \pi_k P_{kj} = \sum_{k=0}^{\infty} \pi_k \sum_{j=0}^{\infty} P_{kj} = \sum_{k=0}^{\infty} \pi_k$$

which is a contradiction, since $\sum_{j=0}^{\infty} \pi_j \leq 1$. Therefore,

$$\pi_j = \sum_{k=0}^{\infty} \pi_k P_{kj}, \qquad j = 0, 1, 2, \ldots$$

Putting $P_j = \pi_j / \sum_0^{\infty} \pi_k$, we see that $\{P_j, j = 0, 1, 2, \ldots\}$ is a stationary distribution, and hence at least one stationary distribution exists.

Let $\{P_j, j = 0, 1, 2, \ldots\}$ be any stationary distribution. Then if $\{P_j, j = 0, 1, 2, \ldots\}$ is the initial probability distribution, we have by (8) that

$$P_j = P\{X_n = j\}$$

$$= \sum_{i=0}^{\infty} P\{X_n = j \mid X_0 = i\} \cdot P\{X_0 = i\}$$

$$= \sum_{i=0}^{\infty} P_{ij}^n P_i \tag{10}$$

Letting $n \to \infty$ in (10) yields

$$P_j = \sum_{i=0}^{\infty} \pi_j P_i = \pi_j$$

and the proof of (b) is complete. If the states are transient or null recurrent and $\{P_j, j = 0, 1, 2, \ldots\}$ is a stationary distribution, then Equations (10) hold and $P_{ij}^n \to 0$, which is clearly impossible. Thus, for Case (a), no stationary distribution exists and the proof is complete.

A similar theorem may be proven in the periodic case. We leave this for the reader.

EXAMPLE 6.   *Positive Recurrent Random Walk.*   Consider a random walk with states $0, 1, 2, \ldots$ for which

$$P_{i, i+1} = p_i, P_{i, i-1} = q_i = 1 - p_i \qquad i = 0, 1, 2, \ldots$$

where $p_0 = 1$. From Theorem 4.9, it follows that this Markov chain is positive recurrent if and only if the system of equations

$$y_0 = y_1 q_1$$

$$y_j = y_{j+1} q_{j+1} + y_{j-1} p_{j-1} \qquad j \geq 1$$

possesses a solution such that $y_j \geq 0$, $\sum_j y_j = 1$. We may now rewrite these equations to obtain

$$y_0 = y_1 q_1$$

$$y_{j+1} q_{j+1} - y_j p_j = y_j q_j - y_{j-1} p_{j-1}, \qquad j \geq 1$$

From the above it follows that

$$y_{j+1} q_{j+1} = y_j p_j, \qquad j \geq 0$$

Hence,

$$y_{j+1} = y_0 \frac{p_0 \cdots \cdots p_j}{q_1 \cdots \cdots q_{j+1}}, \qquad j \geq 0$$

Therefore, a necessary and sufficient condition for the random walk to be positive recurrent is for

$$\sum_{j=0}^{\infty} \frac{p_0 \cdots \cdots p_j}{q_1 \cdots \cdots q_{j+1}} < \infty$$

If $p_j = p$, $q_j = q$ for $j > 0$, then the above reduces to the condition that $p < q$, i.e., that $p < \frac{1}{2}$. The stationary distribution $\{\pi_j, j \geq 0\}$ is given by

$$\pi_j = \frac{y_j}{\sum_{j=0}^{\infty} y_j} = \begin{cases} (q - p)/2q, & j = 0 \\ \dfrac{(q - p)(p/q)^{j-1}}{2q}, & j > 0 \end{cases}$$

The mean recurrence times μ_{jj} are given (when $p < \frac{1}{2}$) by

$$\mu_{jj} = \frac{1}{\pi_j}$$

4.4. Transitions Among Classes

Proposition 4.10

Let R be a recurrent class of states. Then $i \in R$, $j \notin R$ implies that $P_{ij} = 0$.

PROOF. Suppose $P_{ij} > 0$. Then $P_{ji}^n = 0$ for all n (otherwise $i \leftrightarrow j$ and $j \in R$). Thus, if the process starts in i, then there exists positive probability (at least P_{ij}) that the process will not return to i. This contradicts the fact that R is recurrent, hence P_{ij} must be 0.

Thus, if we start in a recurrent class of states, then we never leave this class. For this reason, a recurrent class is said to be a *closed class*.

Let j be a recurrent state. We are often interested in the probability that we will ever enter state j, given that we start in the transient state i. Let T denote the set of transient states.

Proposition 4.11

If j is recurrent, then the set of probabilities $\{f_{ij}, i \in T\}$ satisfy

$$f_{ij} = \sum_{k \in T} P_{ik} f_{kj} + \sum_{k \in R} P_{ik} \qquad i \in T$$

where R denotes the set of states communicating with j.

PROOF.

$$f_{ij} = P\{N_j(\infty) > 0 \,|\, X_0 = i\}$$

$$= \sum_{\text{all } k} P\{N_j(\infty) > 0 \,|\, X_0 = i, X_1 = k\} P\{X_1 = k \,|\, X_0 = i\}$$

$$= \sum_{k \in T} f_{kj} P_{ik} + \sum_{k \in R} f_{kj} P_{ik} + \sum_{\substack{k \notin R \\ k \notin T}} f_{kj} P_{ik}$$

$$= \sum_{k \in T} f_{kj} P_{ik} + \sum_{k \in R} P_{ik}$$

where we have used Corollary 4.6 in asserting that $f_{kj} = 1$ for $k \in R$ and Proposition 4.10 in asserting that $f_{kj} = 0$ for $k \notin T$, $k \notin R$.

EXAMPLE 7. *The Gambler's Ruin.* Consider a gambler who at each play of the game has probability p of winning one unit and probability $q = 1 - p$ of losing one unit. If the successive plays of the game are assumed independent, what is the probability that starting with i units, the gambler's fortune will reach N before reaching 0?

If we let X_n denote the gambler's fortune after the nth play of the game, then the process $\{X_n\}$ is a markov chain with two recurrent classes, namely $\{0\}$ and $\{N\}$. By Proposition 4.11, we obtain

$$f_{0N} = 0$$

$$f_{iN} = p f_{i+1,N} + q f_{i-1,N} \qquad i = 1, 2, \ldots, N-1$$

$$f_{NN} = 1$$

By rearranging these equations and writing f_i for f_{iN}, we obtain

$$f_{i+1} - f_i = \frac{q}{p}(f_i - f_{i-1}) \qquad i = 1, 2, \ldots, N-1$$

Hence,

$$f_2 - f_1 = \frac{q}{p} f_1$$

$$f_3 - f_2 = \frac{q}{p}(f_2 - f_1) = \left(\frac{q}{p}\right)^2 f_1$$

$$\vdots$$

$$1 - f_{N-1} = \left(\frac{q}{p}\right)(f_{N-1} - f_{N-2}) = \left(\frac{q}{p}\right)^{N-1} f_1$$

Adding these equations yields

$$f_i - f_1 = f_1\left[\left(\frac{q}{p}\right) + \left(\frac{q}{p}\right)^2 + \cdots + \left(\frac{q}{p}\right)^{i-1}\right], \qquad i > 1$$

or

$$f_i = \begin{cases} \dfrac{1 - (q/p)^i}{1 - q/p}\, f_1 & \text{if } q/p \neq 1 \\ i f_1 & \text{if } q/p = 1 \end{cases}$$

Finally, using the fact that $f_N = 1$, we obtain

$$f_{iN} = \begin{cases} \dfrac{1 - (q/p)^i}{1 - (q/p)^N} & \text{if } p \neq \tfrac{1}{2} \\ i/N & \text{if } p = \tfrac{1}{2} \end{cases}$$

If we let $N = \infty$, then

$$f_{i\infty} = \begin{cases} 1 - (q/p)^i & p > \tfrac{1}{2} \\ 0 & p \leq \tfrac{1}{2} \end{cases}$$

Thus, if $p > \tfrac{1}{2}$, there is a positive probability of the gambler's fortune increasing indefinitely; and if $p \leq \tfrac{1}{2}$ then with probability one, the gambler will go broke against an infinitely rich adversary.

4.5. Branching Processes

Suppose an organism at the end of its lifetime produces a random number Z of offspring with probability distribution

$$P\{Z = j\} = P_j, \qquad j = 0, 1, 2, \dots \tag{12}$$

where, of course, $P_j \geq 0$ and $\sum_{j=0}^{\infty} P_j = 1$. We shall suppose that all offspring act independently of each other and produce their own offspring in accordance with the probability distribution (12). Under these conditions, it is easy to see that $\{X_n, n = 0, 1, 2, \dots\}$ is a Markov chain, where X_n denotes the population size at the nth generation.

Assume that $X_0 = 1$, that is, the initial population consists of one individual. Clearly, we may write for every $n = 0, 1, 2, \dots$

$$X_{n+1} = \sum_{i=1}^{X_n} Z_i \tag{13}$$

where Z_i, $i \geq 1$ are independent random variables with a common distribution given by (12). Z_i, of course, represents the number of offspring of the ith individual of the nth generation.

From (13), we may easily show that

$$EX_n = m^n \tag{14}$$

and

$$\mathrm{Var}\, X_n = \begin{cases} \sigma^2 m^{n-1} \dfrac{m^n - 1}{m - 1} & \text{if } m \neq 1 \\ n\sigma^2 & \text{if } m = 1 \end{cases} \tag{15}$$

where

$$m = \sum_{j=0}^{\infty} j P_j, \qquad \sigma^2 = \sum_{j=0}^{\infty} (j - m)^2 P_j$$

Let us introduce the generating functions

$$\phi_n(s) = Es^{X_n} = \sum_{j=0}^{\infty} s^j P\{X_n = j\} \qquad n = 1, 2, \ldots$$

From (13), we have

$$\begin{aligned} \phi_{n+1}(s) &= Es^{X_{n+1}} \\ &= E[E(s^{X_{n+1}} \mid X_n)] \\ &= E[E(s^{Z_1 + \cdots + Z_{X_n}} \mid X_n)] \\ &= E[(E[s^{Z_1}])^{X_n}] \\ &= E[\{\phi_1(s)\}^{X_n}] \end{aligned}$$

But the right-hand side is just the generating function $\phi_n(\cdot)$ evaluated at $\phi_1(s)$. Hence,

$$\phi_{n+1}(s) = \phi_n[\phi_1(s)] \tag{16}$$

Now, (16) implies that

$$\phi_{n+1}(s) = \phi_1[\phi_n(s)] \tag{17}$$

This is easily seen by induction as $\phi_2(s) = \phi_1[\phi_1(s)]$ and

$$\phi_{n+1}(s) = \phi_n(\phi_1(s)) = \phi_1(\phi_{n-1}[\phi_1(s)]) = \phi_1[\phi_n(s)]$$

Let π_0 denote the probability that the population will eventually die out; or formally,

$$\pi_0 = \lim_{n \to \infty} P\{X_n = 0\}$$

Note that the limit exists, since $X_n = 0$ implies that $X_{n+1} = 0$, and thus $P\{X_n = 0\}$ is increasing in n.

Theorem 4.12

Suppose that $P_0 > 0$ and $P_0 + P_1 < 1$.
(i) *π_0 is the smallest positive number p satisfying*

$$\phi_1(p) = p$$

(ii) *$\pi_0 = 1$ if and only if $m \leq 1$.*

PROOF. By letting $s = 0$ in (17), we obtain

$$\phi_{n+1}(0) = \phi_1(\phi_n(0))$$

However, by definition of ϕ_n, $\phi_n(0) = P\{X_n = 0\}$. Hence,

$$P\{X_{n+1} = 0\} = \phi_1(P\{X_n = 0\})$$

By letting $n \to \infty$ and using the fact that generating functions are continuous, we obtain

$$\pi_0 = \phi_1(\pi_0)$$

To show that π_0 is the smallest solution, suppose that $p = \phi_1(p)$. Then, using the fact that $\phi_1(s)$ is a nondecreasing function for $s \in [0, 1]$, we have

$$P\{X_1 = 0\} = \phi_1(0) \leq \phi_1(p) = p$$

By induction, we have

$$P\{X_n = 0\} = \phi_n(0) = \phi_1[\phi_{n-1}(0)] \leq \phi_1(p) = p$$

for all $n \geq 1$; and letting $n \to \infty$ yields,

$$\pi_0 \leq p$$

To prove (ii), we first note that if $P_0 + P_1 < 1$, then

$$\phi_1''(s) = \sum_{j=0}^{\infty} j(j-1)s^{j-2}P_j > 0$$

for all $s \in (0, 1)$. Hence, $\phi_1(s)$ is a strictly convex function in the open interval $(0, s)$. We now distinguish two cases (Figures 4.1 and 4.2): In Figure 4.1 $\phi_1(s) > s$ for all $s \in (0, 1)$, and in Figure 4.2, $\phi_1(s) = s$ for some $s \in (0, 1)$. It is geometrically clear that Figure 4.1 represents the appropriate picture when $\phi_1'(1) \leq 1$ and Figure 4.2 is appropriate when $\phi_1'(1) > 1$. Thus, $\pi_0 = 1$ if and only if $\phi_1'(1) \leq 1$. The result follows, since $\phi_1'(1) = \sum_1^{\infty} jP_j = m$.

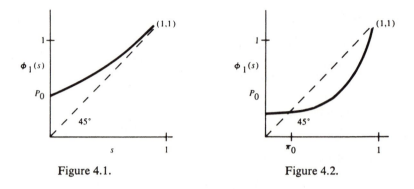

Figure 4.1.   Figure 4.2.

4.6. Transient States

In this section we will derive a necessary and sufficient condition for an irreducible Markov chain to be transient. Let S be a fixed set of states of some irreducible Markov chain $\{X_n\}$, and define for $i \in S$

$$Y_i(n) = P\{X_j \in S \quad \text{for } j = 1, 2, \ldots, n | X_0 = i\}$$

By conditioning on the value of X_1, we obtain

$$Y_i(1) = \sum_{j \in S} P_{ij} \tag{18}$$

and for $n > 1$,

$$Y_i(n) = \sum_{j \in S} P_{ij} Y_j(n-1)$$

Now from its definition it follows that $Y_i(n)$ is a nonincreasing bounded sequence, and hence $Y_i = \lim_{n \to \infty} Y_i(n)$ exists and satisfies

$$Y_i = \sum_{j \in S} P_{ij} Y_j, \quad i \in S$$

Clearly, Y_i denotes the probability that the Markov chain never leaves the set S.

Now, suppose that Z_i, $i \in S$ satisfy

$$Z_i = \sum_{j \in S} P_{ij} Z_j, \quad |Z_i| \leq 1, \quad i \in S \tag{19}$$

From (18) it follows that $|Z_i| \leq Y_i(1)$, and by induction, $|Z_i| \leq Y_i(n)$ for all n. Hence, $|Z_1| \leq Y_i$ for all $i \in S$. Therefore, Y_i, $i \in S$, is the maximal solution of (19) and is thus zero if and only if the system (19) has no nonzero bounded solution.

We shall need the following lemma, whose proof will be left to the reader.

Lemma 4.13

An irreducible Markov chain with state space 0, 1, 2, ... *is recurrent if and only if* $f_{i0} = 1$ *for all* $i \neq 0$.

Theorem 4.14

An irreducible Markov chain with states 0, 1, 2, ... *has all transient states if and only if the system of equations*

$$Z_i = \sum_{j=1}^{\infty} P_{ij} Z_j \qquad i = 1, 2, \ldots \tag{20}$$

has a nonzero bounded solution.

PROOF. Let S consist of the states 1, 2, ... If the system of Equations (20) has a nonzero bounded solution, then $Y_i > 0$ for some i. However, this implies by Lemma 4.13 (since $Y_i = 1 - f_{i0}$) that state 0 is transient. On the other hand, if the system (20) has no nonzero bounded solution, then $Y_i = 0$ for all $i > 0$, and the result follows from the preceding lemma.

EXAMPLE 8. *Random Walk with Variable Parameters.* Consider a random walk with states 0, 1, 2, ... for which

$$P_{i,i+1} = p_i, \, P_{i,i-1} = q_i = 1 - p_i, \qquad i = 0, 1, 2, \ldots$$

where $p_0 = 1$. The system of Equations (20) reduces to

$$Z_1 = p_1 Z_2$$
$$Z_i = p_i Z_{i+1} + q_i Z_{i-1} \qquad i = 2, 3, \ldots$$

Rewriting the above in the form $Z_{i+1} - Z_i = (q_i/p_i)(Z_i - Z_{i-1})$, $i > 1$, and then solving recursively for $Z_{i+1} - Z_i$, yields

$$Z_{i+1} - Z_i = \prod_{j=2}^{i} (q_j/p_j)(Z_2 - Z_1)$$
$$= Z_1 \prod_{j=1}^{i} q_j/p_j, \qquad i \geq 1$$

Thus, by adding the above equations, we obtain

$$Z_{n+1} = Z_1 \sum_{i=0}^{n} \rho_i$$

where $\rho_0 = 1$, $\rho_i = \prod_{j=1}^{i} q_j/p_j$. The system is transient if Z_n is bounded, and hence the system is transient if and only if

$$\sum_{i=0}^{\infty} \rho_i < \infty$$

Therefore, in conjunction with Example 6, we have that the Markov chain is

positive recurrent if $\quad \displaystyle\sum_{j=0}^{\infty} \frac{p_0 \cdots p_j}{q_1 \cdots q_{j+1}} < \infty$

null recurrent if $\quad \displaystyle\sum_{j=0}^{\infty} \frac{p_0 \cdots p_j}{q_1 \cdots q_{j+1}} = \infty \quad$ and $\quad \displaystyle\sum_{j=1}^{\infty} \frac{q_1 \cdots q_j}{p_1 \cdots p_j} = \infty$

transient if $\quad \displaystyle\sum_{j=1}^{\infty} \frac{q_1 \cdots q_j}{p_1 \cdots p_j} < \infty$

EXAMPLE 9. *Classification of States in the* $M/G/1$ *Queueing System.* Consider the Markov chain of Example 3. Its transition probability matrix is of the form

$$P = \begin{Vmatrix} a_0 & a_1 & a_2 & \cdot & \cdot & \cdot \\ a_0 & a_1 & a_2 & \cdot & \cdot & \cdot \\ 0 & a_0 & a_1 & a_2 & \cdot & \cdot \\ 0 & 0 & a_0 & a_1 & \cdot & \cdot \\ \cdot & \cdot & \cdot & \cdot & & \end{Vmatrix}$$

The quantity a_i, $a_i > 0$, $\sum_{i=0}^{\infty} a_i = 1$, represents the probability that i customers arrive during an arbitrary service period for the $M/G/1$ queueing system. Let $\rho = \sum_{u=0}^{\infty} n a_n$. We shall show that the Markov chain is

positive recurrent if $\rho < 1$

null recurrent if $\rho = 1$

transient if $\rho > 1$

We first prove that the chain is positive recurrent when $\rho < 1$ by solving the system of equations $\pi_i = \sum_j P_{ij} \pi_j$, $i \geq 0$, for the stationary probabilities. These equations take the form

$$\pi_i = \pi_0 a_i + \sum_{j=1}^{i+1} \pi_j a_{i-j+1}, \qquad i \geq 0 \tag{21}$$

We solve these equations by introducing the generating functions

$$\pi(s) = \sum_{i=0}^{\infty} \pi_i s^i, \qquad A(s) = \sum_{i=0}^{\infty} a_i s^i$$

Multiplying both sides of (21) by s^i and summing over i yields

$$\pi_0 A(s) - \frac{\pi_0 A(s)}{s} + \frac{A(s)\pi(s)}{s} = \pi(s)$$

or

$$\pi(s) = \frac{\pi_0(s-1)A(s)}{s - A(s)}$$

Now, since $\lim_{s \to 1^-} A(s) = 1$, we obtain

$$\lim_{s \to 1^-} \pi(s) = \pi_0 \lim_{s \to 1^-} \frac{s-1}{s - A(s)} = \pi_0 \lim_{s \to 1^-} \left(1 - \frac{1 - (As)}{1 - s}\right)^{-1}$$

$$= \pi_0 [1 - A'(1)]^{-1}$$

$$= \frac{\pi_0}{1 - \rho}$$

However, since $\lim_{s \to 1^-} \pi(s) = \sum_{i=0}^{\infty} \pi_i$, this implies that $\sum_{i=0}^{\infty} \pi_i = \pi_0/1 - \rho$; thus stationary probabilities exist if and only if $\rho < 1$. Hence, the chain is positive recurrent in this case.

Now, consider the question of transience. In this case, we seek a bounded solution to the Equations (20) which take the form

$$Z_1 = \sum_{j=1}^{\infty} a_j Z_j$$

$$Z_i = \sum_{j=0}^{\infty} a_j Z_{i+j-1}, \quad i > 1 \tag{22}$$

Consider a trial solution, $Z_i = 1 - s^i$. The Equations (22) reduce to the single equation

$$A(s) = s$$

However, from the theory of branching processes we know that this equation has a unique root $s_0 \in (0, 1)$ if and only if $\rho = A'(1) > 1$. Hence, the system is transient if $\rho > 1$.

Finally, it remains to show that the system is null recurrent when $\rho = 1$. To do so, we shall use the following theorem, which we state without proof.

Theorem 4.15

An irreducible aperiodic Markov chain with states $0, 1, 2, \ldots$ is recurrent if there is a solution of the inequalities

$$Z_i \geq \sum_{j=0}^{\infty} P_{ij} Z_j, \quad i = 1, 2, \ldots \tag{23}$$

with the property that $Z_i \to \infty$ as $i \to \infty$.

It is not difficult to show that $Z_j = j$ satisfies the inequality (23) if $\rho = 1$, and hence the system is recurrent in this case. Since we have previously shown that the Markov chain is positive recurrent only when $\rho < 1$, the result follows.

Note that our conclusions are rather intuitive. Since ρ is the mean number of customers arriving during a busy period, it follows that when $\rho > 1$, there are, on the average, more people arriving than departing in each period. Hence it seems reasonable that the queue size should increase indefinitely. On the other hand, if $\rho < 1$, then the mean number of arrivals per period is less than the mean number of departures and hence it is not unreasonable that the queue size should have a limiting distribution.

Also, it should be noted that the condition that $\rho < 1$ is equivalent to

$$\frac{\text{Mean service time}}{\text{Mean interarrival time}} < 1$$

Problems

1. N white and N black balls are distributed in two urns in such a way that each contains N balls. We say that the system is in state i ($i = 0, 1, \ldots, N$) if the first urn contains i black balls. At each step, we draw one ball from each urn and place the ball drawn from the second urn into the first urn, and conversely with the ball from the first urn. Determine the transition probabilities P_{ij}.

2. Specify the classes of the following Markov chains, give their period and determine whether or not they are recurrent.

$$P_1 = \begin{Vmatrix} 0 & \frac{1}{2} & \frac{1}{2} \\ \frac{1}{2} & 0 & \frac{1}{2} \\ \frac{1}{2} & \frac{1}{2} & 0 \end{Vmatrix}, \qquad P_2 = \begin{Vmatrix} 0 & 0 & \frac{1}{2} & \frac{1}{2} \\ 1 & 0 & 0 & 0 \\ 0 & 1 & 0 & 0 \\ 0 & 1 & 0 & 0 \end{Vmatrix}$$

$$P_3 = \begin{Vmatrix} \frac{1}{2} & \frac{1}{2} & 0 & 0 & 0 \\ \frac{1}{2} & \frac{1}{2} & 0 & 0 & 0 \\ 0 & 0 & \frac{1}{2} & \frac{1}{2} & 0 \\ 0 & 0 & \frac{1}{2} & \frac{1}{2} & 0 \\ \frac{1}{4} & \frac{1}{4} & 0 & 0 & \frac{1}{2} \end{Vmatrix}$$

3. Consider a Markov chain with states $0, 1, 2, \ldots$ and with transition probabilities given by

$$P_{0i} = P_i > 0, \qquad \sum_{i=0}^{\infty} P_i = 1, \qquad \sum_{i=0}^{\infty} i P_i < \infty$$

$$P_{i,i-1} = 1 \qquad\qquad (i \geq 1)$$

Show that the chain is irreducible, aperiodic, recurrent and positive recurrent, and find the stationary probability distribution.

4. Consider the generating functions $F_{ij}(s) = \sum_{n=0}^{\infty} f_{ij}^n s^n$ and $P_{ij}(s) = \sum_{n=0}^{\infty} P_{ij}^n s^n$. Show that

$$P_{ij}(s) = P_{jj}(s)F_{ij}(s) \qquad \text{when } i \neq j$$

$$P_{ii}(s) = \frac{1}{1 - F_{ii}(s)}$$

Use this to give another proof of Proposition 4.4.

5. A transition matrix P is said to be doubly stochastic if the sum over each column equals one, i.e.,

$$\sum_i P_{ij} = 1 \qquad \text{for all } j$$

If such a chain is irreducible and aperiodic, and consists of $M(M < \infty)$ states, calculate the limiting probabilities.

6. Show that positive and null recurrence are class properties.

7. Use Wald's equation to show that the random walk with $P_{i,i+1} = \frac{1}{2} = P_{i,i-1}$ $i = 0, \pm 1, \ldots$, is null recurrent.

8. Suppose that if it rains today, then it will rain tomorrow with probability α, while if it does not rain today, then it will rain tomorrow with probability β. Calculate the limiting probability of rain.

9. Let X_n be the sum of n independent rolls of a fair die. Find

$$\lim_n P\{X_n \text{ is a multiple of } 13\}$$

10. If $f_{jj} < 1$, show that

(i) $\sum_{n=1}^{\infty} P_{ij}^n < \infty$

(ii) $\lim_{n \to \infty} P_{ij}(n) = 0$

(iii) $\sum_{n=1}^{\infty} P_{jj}^n = \dfrac{f_{jj}}{1 - f_{jj}}$

11. Show that for i and j transient,

$$f_{ij} = \frac{\sum_{n=1}^{\infty} P_{ij}^n}{1 + \sum_{n=1}^{\infty} P_{jj}^n}$$

Hint: Use the fact that $E[N_j(\infty)|X_0 = i] = \sum_{n=1}^{\infty} P_{ij}^n$.

12. Prove that in a finite Markov chain, there are no null recurrent states and not all states can be transient.

13. Let T denote the transient states. For $i \in T$, let

$$M_i = E[\text{time until chain enters a recurrent state} \,|X_0 = i]$$

Show that $M_i = 1 + \sum_{j \in T} P_{ij} M_j$, $i \in T$

14. Find M_i, $i = 1, 2, \ldots, N-1$, for the gambler's ruin problem of Example 7. Hint: A simple way of doing this is to use Wald's equation in conjunction with the quantities f_{iN}.

15. Prove Equations (14) and (15).
16. In a branching process, suppose $\lim_{n \to \infty} P\{X_n = 0 | X_0 = 1\} = \frac{1}{3}$. Then,

$$\lim_{n \to \infty} P\{X_n = 0 | X_0 = j\} = ?$$

17. Consider an aperiodic, irreducible Markov chain having stationary probabilities π_i. Suppose that X_0 is determined according to the π_i, and let

$$Q_{ij} = P\{X_0 = j | X_1 = i\}$$

Then $Q = \|Q_{ij}\|$ may itself be considered as a transition matrix for some Markov chain. Show that

$$Q_{ij}^n = P\{X_0 = j | X_n = i\}$$

18. For the embedded chain of the $G/M/1$ queueing system (see Example 4), let

$$a_i = \int_0^\infty e^{-\lambda x} \frac{(\lambda x)^i}{i!} \, dG(x), \qquad i \geq 0$$

Show that there exists a stationary distribution if $\sum_i i a_i > 1$.
Hint: Consider solutions of the equations $\pi_i = \sum_j P_{ij} \pi_j$ of the form $\pi_i = s^i$.
19. Show that $Z_j = j$ satisfies the inequality (23).
20. In Example 9, show that $\rho < 1$ is equivalent to

$$\frac{\text{Mean service time}}{\text{Mean interarrival time}} < 1$$

References

The renewal theoretic approach towards Markov chains, which we have adopted, is due to Feller. For further results on branching processes, the interested reader should consult Harris [3].

[1] Cox, D. R. and H. D. Miller. *The Theory of Stochastic Processes*, John Wiley and Sons, New York, (1965).

[2] Feller, W. *An Introduction to Probability Theory and its Applications*, Vol. 1, John Wiley and Sons, New York, (1957).

[3] Harris, T. *The Theory of Branching Processes*, Springer-Verlag, Berlin, (1963).

[4] Karlin, S. *A First Course in Stochastic Processes*, Academic Press, New York, (1966).

[5] Kemeny, J., J. L. Snell and A. Knapp. *Denumerable Markov Chains*, Van Nostrand, New Jersey, (1966).

[6] Parzen, E. *Stochastic Processes*, Holden-Day, San Francisco, (1962).

[7] Prabhu, N. U. *Stochastic Processes*, MacMillan, New York, (1965).

[8] Smith, W. "Renewal Theory and Its Ramifications," *Journal of the Royal Statistical Society*, Series B., **20**, pp. 243–302, (1958).

5

SEMI-MARKOV, MARKOV RENEWAL AND REGENERATIVE PROCESSES

5.1. Introduction and Preliminaries

In this chapter we shall consider a stochastic process which makes transitions from state to state in accordance with a Markov chain, but in which the amount of time spent in each state before a transition occurs is random.

Unless otherwise mentioned, we shall denote the state space by the non-negative integers $0, 1, 2, \ldots$, and we let $Q_{ij}(t)$ denote the probability that after making a transition into state i, the process next makes a transition into state j, in an amount of time less than or equal to t. Clearly, we must have

$$Q_{ij}(t) \geq 0, \qquad i, j = 0, 1, 2, \ldots, \quad t \geq 0$$

$$\sum_{j=0}^{\infty} Q_{ij}(\infty) = 1, \qquad i = 0, 1, 2, \ldots$$

Let

$$P_{ij} = Q_{ij}(\infty)$$

and if $P_{ij} > 0$, let

$$F_{ij}(t) = \frac{Q_{ij}(t)}{P_{ij}}$$

If $P_{ij} = 0$, let $F_{ij}(t)$ be arbitrary.

Hence, P_{ij} represents the probability that the next transition will be into state j, given that the process has just entered i; and $F_{ij}(t)$ represents the conditional probability that a transition will take place within an amount of time t, given that the process has just entered i and will next enter j. Thus, transitions may be thought of as taking place in two stages. When i is entered, the next state is chosen according to the transition probabilities P_{ij}; then given that the state chosen is j, the time until transition has a distribution $F_{ij}(\cdot)$.

Let J_0 denote the initial state of the process; and for $n \geq 1$, let J_n denote the state of the process immediately after the nth transition has occurred. It is clear that the process $\{J_n, n = 0, 1, 2, \ldots\}$ is a Markov Chain with transition probabilities P_{ij}. We shall refer to this process as the *embedded Markov Chain*.

Also, let $N_i(t)$ denote the number of transitions into state i which occur in $(0, t]$; and define

$$N(t) = \sum_{i=0}^{\infty} N_i(t)$$

If we let $Z(t)$ denote the state of the process at time t, then

$$Z(t) = J_{N(t)}$$

Definition 5.1

The stochastic process $\{Z(t), t \geq 0\}$ is called a *Semi-Markov Process*.

Definition 5.2

Setting $N(t) = (N_1(t), N_2(t), \ldots)$, the stochastic process $\{N(t), t \geq 0\}$ is called a *Markov Renewal Process*.

Thus, the semi-Markov process records the state of the process at each time point, while the Markov renewal process is a counting process which keeps track of the number of times each state has been visited.

It is easy to show (see Problem 1) that a knowledge of the initial state J_0 and the Markov renewal process $\{N(t), t \geq 0\}$ determines the semi-Markov process $\{Z(t), t \geq 0\}$. On the other hand, $\{Z(t), t \geq 0\}$ determines $\{N(t), t \geq 0\}$ only when $P_{ii} = 0$ for all i. To see this, suppose that there is only one state. Then, $\{N(t), t \geq 0\}$ is a renewal process with renewal distribution $Q_{00}(\cdot)$, while $\{Z(t), t \geq 0\}$ is the trivial process which is identically 0. As a matter of fact, one of the major reasons we allow transitions from a state to itself is to include renewal processes as a special kind of Markov renewal processes.

Let $H_i(\cdot)$ denote the distribution at the amount of time until the next transition occurs, given that the process has just entered i. That is,

$$H_i(t) = \sum_{j=0}^{\infty} P_{ij} F_{ij}(t) = \sum_{j=0}^{\infty} Q_{ij}(t)$$

We suppose that $H_i(0) < 1$ for all i.

The first question we shall consider is whether or not an infinite number of transitions can occur within a finite amount of time.

Definition 5.3

We say that state i is *regular* if

$$P\{N(t) = \infty \mid J_0 = i\} = 0 \qquad \text{for all } t < \infty$$

We say that the Markov renewal process is regular if all the states are regular. Suppose that the initial state is j. Then it is easy to see that transitions into j constitute renewals. That is, the successive times between transitions into j are independent and identically distributed. Thus, for any initial state i, $\{N_j(t), t \ge 0\}$ constitutes a delayed (possibly terminating) renewal process in which the initial distribution need not be honest. Hence, from renewal theory it follows that with probability 1,

$$N_j(t) < \infty \qquad \text{for all } t$$

Thus, if the number of states is finite, then with probability 1,

$$N(t) < \infty \qquad \text{for all } t \ge 0$$

Hence, a finite state Markov renewal process is necessarily regular. However, when the state space is countable, certain conditions are needed to ensure regularity. For example, suppose that

$$P_{ii+1} = 1 \qquad i = 0, 1, 2, \ldots$$

and

$$F_{ii+1}(t) = \begin{cases} 0 & t < (\tfrac{1}{2})^i \\ 1 & t \ge (\tfrac{1}{2})^i \end{cases}$$

Or, in other words, the process goes from i to $i + 1$ in an amount time $(\tfrac{1}{2})^i$. It then follows that

$$P\{N(2) = \infty \mid J_0 = i\} = 1, \qquad i = 0, 1, 2, \ldots$$

and hence no state is regular.

Before giving sufficient conditions for regularity, it will be convenient to denote by X_{n+1} the elapsed time between the nth and $(n + 1)$th transition. It should be clear that the process $\{(J_n, X_{n+1}), n = 0, 1, 2, \ldots\}$ completely determines the Markov renewal process when this counting process is regular. Also, it follows that

$$N(t) = \sup\{n : X_1 + \cdots + X_n \le t\} \qquad (1)$$

Proposition 5.1

If

(a) there exists $\alpha > 0$, $\varepsilon > 0$, such that

$$1 - H_i(\alpha) > \varepsilon, \qquad \text{for all } i = 0, 1, 2, \ldots \tag{2}$$

or if

(b) for each initial state $J_0 = i$, the embedded Markov chain

$$\{J_n, n = 0, 1, 2, \ldots\}$$

will, with probability 1, eventually reach a recurrent state

then the Markov renewal process is regular.

PROOF. To prove (a), define $\overline{X}_n$, $n = 1, 2, \ldots$, by

$$\overline{X}_n = \begin{cases} 0 \text{ if } X_n \leq \alpha \\ \alpha \text{ with probability } \dfrac{\varepsilon}{1 - H_j(\alpha)} & \text{if } X_n > \alpha, J_{n-1} = j \\ 0 \text{ with probability } 1 - \dfrac{\varepsilon}{1 - H_j(\alpha)} & \text{if } X_n > \alpha, J_{n-1} = j \end{cases}$$

It follows from (2) that $\overline{X}_n$ is well defined, and also that $\overline{X}_n$, $n = 1, 2, \ldots$, are independent and mutually distributed random variables, having

$$P\{\overline{X}_n = \alpha\} = \varepsilon = 1 - P\{\overline{X}_n = 0\}$$

Hence, from renewal theory,

$$\overline{N}(t) = \sup\{n : \overline{X}_1 + \cdots + \overline{X}_n \leq t\} < \infty \quad \text{with probability 1}$$

and the result follows from (1), since $\overline{X}_n \leq X_n$ for all n.

To prove the sufficiency of (b), let j be the first recurrent state that is reached, and suppose it was reached at the n_0th transition (n_0 must be finite by assumption). Let $n_1, n_2, \ldots$ be the successive integers n at which $J_n = j$. (Such integers exist since j is recurrent.) Set $T_0 = X_1 + \cdots + X_{n_0}$, and

$$T_k = X_{n_{k-1}+1} + \cdots + X_{n_k}.$$

In other words, T_k denotes the amount of time between the kth and $(k + 1)$th visit to j. Therefore, it follows that $\{T_k, k \geq 1\}$ forms a renewal process, and so $\sum_{k=1}^{\infty} T_k = \infty$ (with probability 1). Since

$$\sum_{n=1}^{\infty} X_n = \sum_{k=0}^{\infty} T_k$$

it follows that

$$\sum_{n=1}^{\infty} X_n = \infty$$

and so the result follows from (1).

We shall assume from here on that all processes considered are regular.

EXAMPLE.   *The* M/M/1 *Queue.*   Consider a single-server queueing system in which customers arrive at a Poisson rate λ and in which the service times are independent and identically distributed exponential random variables with mean $1/\mu$. If we let $n(t)$ denote the number of customers in the system at time t, then it is easy to see that $\{n(t), t \geq 0\}$ is a semi-Markov process with

$$Q_{01}(t) = 1 - e^{-\lambda t}$$

$$Q_{i, i-1}(t) = \frac{\mu}{\lambda + \mu}(1 - e^{-(\lambda + \mu)t}), \qquad i \geq 1$$

$$Q_{i, i+1}(t) = \frac{\lambda}{\lambda + \mu}(1 - e^{-(\lambda + \mu)t}), \qquad i \geq 1$$

EXAMPLE.   *The* M/G/1 *Queue.*   Consider the previous example, but suppose now that the service time distribution is arbitrary. In this case, it is not true that $\{n(t), t \geq 0\}$ is a semi-Markov process. This is so since the state of the system after a transition depends not only on $n(t)$, but also on the amount of time that the person presently being served has been receiving service. However, if we view a *transition* as taking place only when a customer finishes service, then the above is a semi-Markov process. That is, if we say that $Z(t) = n$ if the last time (prior to t) that a customer completed service there were n customers left in the system, then $\{Z(t), t \geq 0\}$ is a semi-Markov process; and it is often called the embedded semi-Markov process of the M/G/1 queue.

5.2. Classification of States

In this section, the states of a Markov renewal process will be classified in much that same manner as is done for Markov chains, the terminology of the latter being retained. Let us define, for all i, j and $t \geq 0$,

$$P_{ij}(t) = P\{Z(t) = j \mid Z(0) = i\} \tag{3}$$

$$G_{ij}(t) = P\{N_j(t) > 0 \mid Z(0) = i\} \tag{4}$$

Hence, $G_{ij}(\cdot)$ is the distribution function of the time until the first transition into j, given that the process starts in i. Note that $G_{ij}(\infty)$ represents the probability that a Markov renewal process starting in i will ever make a transition into j.

Define the moments μ_{ij} by

$$\mu_{ij} = \begin{cases} \infty & \text{if} \quad G_{ij}(\infty) < 1 \\ \int_0^\infty t \, dG_{ij}(t) & \text{otherwise} \end{cases}$$

μ_{ii} is called the *mean recurrence time* of state i.

Let us also denote by μ_i the expected amount of time spent in i during each visit, and by η_{ij} the expected amount of time spent in i during each visit, given that the next state entered is j. That is,

$$\mu_i = \int_0^\infty t \, dH_i(t)$$

and

$$\eta_{ij} = \int_0^\infty t \, dF_{ij}(t)$$

It easily follows that

$$\mu_i = \sum_{j=0}^\infty P_{ij} \eta_{ij} \tag{5}$$

Definition 5.4

(a) State i and j are said to communicate if either $i = j$ or

$$G_{ij}(\infty) \cdot G_{ji}(\infty) > 0.$$

(b) Communication is an equivalence relation, and the disjoint equivalence classes are simply called *classes* and are denoted by C_i (whenever $i \in C_i$).

(c) A Markov renewal (semi-Markov) process is said to be irreducible if there is only one class.

(d) State i is said to be recurrent if $G_{ii}(\infty) = 1$, transient otherwise.

(e) State i is said to be positive recurrent if it is recurrent and $\mu_{ii} < \infty$. State i is said to be null recurrent if it is recurrent and $\mu_{ii} = \infty$.

As might be expected, the properties defined here for Markov renewal processes are very closely related to those of the embedded Markov chain $\{J_n, n \geq 0\}$. This is illustrated by the following proposition.

Proposition 5.2

For a given Markov renewal process, state i is recurrent (is transient) [communicates with state j] if, and only if, state i is recurrent (is transient) [communicates with state j] in the embedded Markov chain.

PROOF. From (4) and the assumed regularity of the process, it follows that

$$G_{ij}(\infty) = P\{J_n = j \quad \text{for some} \quad n > 0 \,|\, J_0 = i\} \tag{6}$$

This relation suffices to verify the proposition, since the properties of recurrence, transience and communication involve only the quantities $G_{ij}(\infty)$, and since (6) shows that these quantities are identical to the analogous quantities of the embedded Markov chain.

For the property of positive (null) recurrence, however, things are not quite as straightforward. Examples may readily be constructed to show that a state of a Markov Renewal process may be positive (null) recurrent, while the same state in the embedded Markov chain is null (positive) recurrent. We can, however, prove the following theorem.

Theorem 5.3

If the state space is finite, then state i is positive recurrent if and only if it is positive recurrent in the embedded Markov chain, and $\mu_j < \infty$ for all $j \in C_i$.

PROOF. Let δ_n denote the set of sequences of states $(i_0, i_1, \ldots, i_n)$ such that

$$i_0 = i_n = i$$
$$i_j \neq i \qquad j = 1, 2, \ldots, n - 1$$

Then it easily follows that

$$\mu_{ii} = \sum_{n=1}^{\infty} \sum_{(i_0, i_1, \cdots, i_n) \in \delta_n} \prod_{k=0}^{n-1} P_{i_k i_{k+1}} (\eta_{i_0 i_1} + \eta_{i_1 i_2} + \cdots + \eta_{i_{n-1} i_n}) \tag{7}$$

Thus, using the fact that the process cannot leave the recurrent class C_i, we have from (7) that

$$\left(\min_{j,\, k \in C_i} \eta_{jk} \right) \mu_{ii}^* \leq \mu_{ii} \leq \left(\max_{j,\, k \in C_i} \eta_{jk} \right) \mu_{ii}^* \tag{8}$$

where

$$\mu_{ii}^* = \sum_{n=1}^{\infty} n \sum_{(i_0, i_1, \ldots, i_n) \in \delta_n} \prod_{k=0}^{n-1} P_{i_k i_{k+1}}$$

However, it is easily seen that μ_{ii}^* is just the mean recurrence time of i in the embedded Markov chain. Now, if i is positive recurrent in the embedded chain (that is, if $\mu_{ii}^* < \infty$), and if $\mu_j < \infty$ for all $j \in C_i$, then it follows that $\eta_{jk} < \infty$ for all $j, k \in C_i$, and hence from (8) that $\mu_{ii} < \infty$. On the other hand, if $\mu_{ii} < \infty$, then it follows that $\mu_j < \infty$ for all $j \in C_i$ (prove this), which implies that $\eta_{jk} < \infty$ for all $j, k \in C_i$, and hence from (8) that $\mu_{ii}^* < \infty$.

From Proposition 5.2 it follows that state j is recurrent if and only if $\sum_{n=1}^{\infty} P_{jj}^n = \infty$. The following criterion is also useful.

Theorem 5.4

If $\mu_j < \infty$, *then j is recurrent if and only if*

$$\int_0^{\infty} P_{jj}(t)\, dt = \infty$$

PROOF.   Suppose $J_0 = j$. By viewing transitions into j as renewals, the question of recurrence becomes one of whether or not the renewal distribution $G_{jj}(\cdot)$ is honest. Hence, by Proposition 3.18, it follows that j is recurrent if and only if

$$EN_j(\infty) = \infty$$

Now, letting

$$A(t) = \begin{cases} 1 & \text{if } Z(t) = j \\ 0 & \text{if } Z(t) \neq j \end{cases}$$

we have that

$$A = \int_0^{\infty} A(t)\, dt$$

represents the total amount of time that the process spends in j. Also,

$$EA = E \int_0^{\infty} A(t)\, dt = \int_0^{\infty} EA(t)\, dt = \int_0^{\infty} P_{jj}(t)\, dt \tag{9}$$

However,

$$A = Y_0 + \sum_{n=1}^{N_j(\infty)} Y_n$$

where Y_n represents the amount of time spent in state j during the nth visit, $n \geq 0$. Hence, by Wald's equation (Theorem 3.6),

$$\begin{aligned} EA &= EY_0 + EY_1 EN_j(\infty) \\ &= \mu_j + \mu_j EN_j(\infty) \end{aligned} \tag{10}$$

{Since $\mu_j < \infty$, a simple truncation argument [on $N_j(\infty)$] shows that (10) holds even when $EN_j(\infty) = \infty$.} Thus, $EN_j(\infty) = \infty$ if and only if $EA = \infty$, and the result follows from (9).

REMARK. If j is transient, then $N_j(\infty)$ is a geometric random variable with mean $G_{jj}(\infty)/1 - G_{jj}(\infty)$; hence from (9) and (10) we obtain

$$\int_0^\infty P_{jj}(t)\, dt = \mu_j \left(1 + \frac{G_{jj}(\infty)}{1 - G_{jj}(\infty)} \right)$$

$$= \frac{\mu_j}{1 - G_{jj}(\infty)} \tag{11}$$

If either $\mu_j = \infty$ or j is recurrent, then both sides of (11) are infinite; thus (11) is always true.

5.3. Some Simple Relationships

Let us begin this section by determining some simple relationships between $P_{ij}(t)$, $G_{ij}(t)$ and $Q_{ij}(t)$. First of all, the relationship between $P_{ij}(t)$ and $Q_{ij}(t)$ is given by

Proposition 5.5

For all states i, j and $t \geq 0$,

$$P_{jj}(t) = 1 - \sum_{k=0}^\infty \int_0^t (1 - P_{kj}(t - x))\, dQ_{jk}(x)$$

$$P_{ij}(t) = \sum_{k=0}^\infty \int_0^t P_{kj}(t - x)\, dQ_{ik}(x), \qquad i \neq j$$

PROOF. By conditioning on both the next state entered and the time at which it is entered, we obtain

$$P_{jj}(t) = \sum_{k=0}^\infty \int_0^\infty P\{Z(t) = j \mid J_0 = j, J_1 = k, X_1 = x\}\, dQ_{jk}(x)$$

However,

$$P\{Z(t) = j \mid J_0 = j, J_1 = k, X_1 = x\} = \begin{cases} 1 & \text{if } x \geq t \\ P_{kj}(t - x) & \text{if } x < t \end{cases}$$

Hence,

$$P_{jj}(t) = \sum_{k=0}^\infty \int_0^t P_{kj}(t - x)\, dQ_{jk}(x) + \sum_{k=0}^\infty \int_t^\infty dQ_{jk}(x)$$

and the result follows, since

$$1 = \sum_{k=0}^{\infty} \int_0^{\infty} dQ_{jk}(x)$$

The case $i \neq j$ is handled in a similar manner.

The analogous relationship between $G_{ij}(t)$ and $Q_{ij}(t)$ is given in

Proposition 5.6

For all states $i, j, t \geq 0$,

$$G_{ij}(t) = \sum_{k=0}^{\infty} \int_0^t G_{kj}(t - x) \, dQ_{ik}(x) + \int_0^t (1 - G_{jj}[t - x]) \, dQ_{ij}(x)$$

The proof of this, which is similar to that of the previous proposition, will be left to the reader.

Finally, the relationship between $P_{ij}(t)$ and $G_{ij}(t)$ is given in

Proposition 5.7

For all states i, j and $t \geq 0$,

$$P_{jj}(t) = \int_0^t P_{jj}(t - x) \, dG_{jj}(x) + 1 - H_j(t)$$

$$P_{ij}(t) = \int_0^t P_{jj}(t - x) \, dG_{ij}(x) \qquad i \neq j$$

PROOF. It suffices to remark that for $i \neq j$, $P_{ij}(t)$ is the probability of reaching state j for the first time before t [according to $G_{ij}(\cdot)$] and then, in the remaining time, ending up in state j [according to $P_{jj}(\cdot)$]. In case $i = j$, one must add to the above the probability that no transition occurs in $(0, t]$, namely, $1 - H_j(t)$.

Let us now consider the random variable $N_j(t)$. Since $\{N_j(t), t \geq 0\}$ is a (possibly delayed) renewal process, it follows [see (3.7) of Chapter 3] that

$$P\{N_j(t) = n \,|\, J_0 = i\} = G_{ij} * (G_{jj})_{n-1}(t) - G_{ij} * (G_{jj})_n(t)$$

where $(G_{jj})_n(\cdot)$ represents the nth fold convolution of $G_{jj}(\cdot)$. Also, since j recurrent and $i \in C_j$ implies that $G_{ij}(\infty) = 1$, we have

$$P\left\{ \lim_{t \to \infty} \frac{N_j(t)}{t} = \frac{1}{\mu_{jj}} \,\middle|\, J_0 = i \right\} = 1 \qquad \text{if } i \in C_j$$

Also, letting

$$m_{ij}(t) = E[N_j(t) \,|\, J_0 = i]$$

it follows from the elementary renewal theorem that

$$\frac{m_{ij}(t)}{t} \to \frac{G_{ij}(\infty)}{\mu_{jj}} \qquad \text{as} \quad t \to \infty$$

and from Blackwell's theorem, that if $G_{jj}(\cdot)$ is not lattice, then for all $a \geq 0$,

$$m_{ij}(t + a) - m_{ij}(t) \to G_{ij}(\infty)\frac{a}{\mu_{jj}} \qquad \text{as} \quad t \to \infty$$

We are now ready to consider the question of a limiting distribution. However, before doing so, we shall first consider a class of stochastic processes known as *regenerative processes*.

5.4. Regenerative Processes

Consider a stochastic process $\{X(t), t \geq 0\}$ with state space $\{0, 1, 2, \dots\}$, having the property that there exist time points at which the process (probabilistically) restarts itself. That is, suppose that with probability one, there exists a time T_1, such that the continuation of the process beyond T_1 is a probabilistic replica of the whole process starting at 0. Note that this property implies the existence of further times $T_2, T_3, \dots$, having the same property as T_1. Such a stochastic process is known as a *regenerative process*.

From the above, it follows that $\{T_1, T_2, \dots\}$ forms a renewal process; and we shall say that a cycle is completed every time a renewal occurs.

EXAMPLES.   (1) A renewal process is regenerative, and T_1 represents the time of the first renewal; (2) a recurrent Markov renewal process is regenerative, and T_1 represents the time of the first transition into the initial state.

The proof of the following important theorem is a further indication of the power of the key renewal theorem.

Theorem 5.8

If T_1 has an absolutely continuous component† and $ET_1 < \infty$, then

$$P_j = \lim_{t \to \infty} P\{X(t) = j\} = E\frac{[\text{amount of time in state } j \text{ during one cycle}]}{E[\text{time of one cycle}]}$$

for all $j \geq 0$.

† That is, it has a density on some interval.

PROOF. Let $P_j(t) = P\{X(t) = j\}$, and denote the distribution function of T_1 by $F(\cdot)$. Then, by conditioning on T_1, we obtain

$$P_j(t) = \int_0^\infty P\{X(t) = j \mid T_1 = s\}\, dF(s)$$

$$= \int_0^t P_j(t - s)\, dF(s) + \int_t^\infty P\{X(t) = j \mid T_1 = s\}\, dF(s) \tag{12}$$

Letting $q_j(t) = \int_t^\infty P\{X(t) = j \mid T_1 = s\}\, dF(s)$, we have, by Proposition 3.4, that the solution of the renewal-type equation (12) is

$$P_j(t) = q_j(t) + \int_0^t q_j(t - s)\, dm(s)$$

and hence from the key renewal theorem,

$$\lim_{t \to \infty} P_j(t) = \frac{\int_0^\infty q_j(t)\, dt}{ET_1} \tag{13}$$

(It is possible to show that $q_j(t)$ is directly Riemann integrable.) Now,

$$P\{X(t) = j,\, T_1 > t\} = \int_0^\infty P\{X(t) = j,\, T_1 > t \mid T_1 = s\}\, dF(s)$$

$$= \int_t^\infty P\{X(t) = j \mid T_1 = s\}\, dF(s)$$

$$= q_j(t)$$

Hence, from (13),

$$\lim_{t \to \infty} P_j(t) = \frac{\int_0^\infty P\{X(t) = j,\, T_1 > t\}\, dt}{ET_1} \tag{14}$$

Now, if we let

$$Y(t) = \begin{cases} 1 & \text{if } X(t) = j \text{ and } T_1 > t \\ 0 & \text{otherwise} \end{cases}$$

then $\int_0^\infty Y(t)\, dt$ represents the amount of time the process is in state j during one cycle. The result then follows from (14), since

$$E \int_0^\infty Y(t)\, dt = \int_0^\infty E[Y(t)]\, dt = \int_0^\infty P\{X(t) = j,\, T_1 > t\}\, dt$$

AN INVENTORY EXAMPLE. Consider an inventory model in which demands for a commodity come in at the end of each day. The successive demands are assumed independent and identically distributed, according to a discrete distribution F. The following (s, S) ordering policy is used: If the inventory level at the beginning of a day is less than or equal to s, then we order enough to bring it up to S, while if the level is greater than s, then we do not order. The order is assumed to be instantaneously filled.

Let X_n be the inventory level at the beginning of the nth day, just after the order (if any) is delivered. If $X_1 = S$, then it is easily seen that $\{X_n, n \geq 1\}$ is a discrete time regenerative process with regeneration point T, the time at which the first order is placed. Hence, by Theorem 5.8, we have that

$$\lim_{n \to \infty} P\{X_n \geq j\} = \frac{E[\text{amount of time the inventory } \geq j \text{ in a cycle}]}{ET}$$

Now, if we let

$$N = \min \{n : Y_1 + \cdots + Y_n > S - j\}$$

where $Y_1, Y_2, \ldots$ are the successive demands, then it follows that the amount of time during the cycle that the inventory level is at least j, $s \leq j \leq S$, is N. Similarly, it follows that T, the length of the cycle, is given by

$$T = \min \{n : Y_1 + \cdots + Y_n > S - s\}$$

However, since $Y_1, Y_2, \ldots$ are independent and identically distributed, it follows that $N - 1(T - 1)$ is just the number of renewals by time $S - j(S - s)$ of the renewal process with interarrival times $Y_1, Y_2, \ldots$. Hence,

$$EN = m(S - j) + 1$$

and

$$ET = m(S - s) + 1$$

where

$$m(t) = \sum_{n=1}^{\infty} F_n(t)$$

Thus, we have shown that

$$\lim_{n \to \infty} P\{X_n \geq j\} = \begin{cases} 1 & j < s \\ \dfrac{m(S - j) + 1}{m(S - s) + 1} & s \leq j \leq S \\ 0 & j > S \end{cases}$$

and the example is completed.

Let us now impose a *reward* structure on the process in the following manner. Suppose that when the process is in state j, we earn a reward at a rate $f(j)$, $j \geq 0$. However, because of the regenerative nature of the process, it follows that what we now have is a renewal reward process, and thus from Proposition 3.16 we obtain

Proposition 5.9

If $E|\int_0^{T_1} f(X_t) \, dt|$ and ET_1 are finite, then
(a) with probability 1,

$$\int_0^t f(X_s) \, ds/t \to \frac{E \int_0^{T_1} f(X_s) \, ds}{ET_1} \qquad \text{as } t \to \infty$$

(b) $E \int_0^t f(X_s) \, ds/t \to \dfrac{E \int_0^{T_1} f(X_s) \, ds}{ET_1} \qquad \text{as } t \to \infty$

While Proposition 5.9 gives us an expression for the long-run (expected) average return, it would be more pleasing if this expression was in terms of the limiting probabilities $P_j, j \geq 0$. For instance, it is somewhat intuitive that the limiting average return should just be the mean reward rate with respect to the limiting distribution, that is, $\sum_{j=0}^{\infty} P_j f(j)$. That this is indeed the case will now be proven.

Theorem 5.10

If the conditions of Theorem 5.8 hold, and if $E|\int_0^{T_1} f(X_t) \, dt|$ and ET_1 are finite, then
(a) *with probability* 1,

$$\int_0^t f(X_s) \, ds/t \to \sum_{j=0}^{\infty} P_j f(j) \qquad \text{as } t \to \infty$$

(b) $E \int_0^t f(X_s) \, ds/t \to \sum_{j=0}^{\infty} P_j f(j) \qquad \text{as } t \to \infty.$

PROOF.   By the definition of the integral, we have

$$\int_0^{T_1} f(X_t) \, dt = \sum_{j=0}^{\infty} f(j) \cdot [\text{amount of time in state } j \text{ during one cycle}]$$

Hence,

$$E \int_0^{T_1} f(X_t) \, dt = \sum_{j=0}^{\infty} f(j) E[\text{amount of time in state } j \text{ during one cycle}]$$

$$= \sum_{j=0}^{\infty} f(j) P_j ET_1$$

where the last equality follows by Theorem 5.8. The result then follows from Proposition 5.9.

As an application of this theorem, we shall now prove that the limiting probability P_j also represents the (long-run) proportion of time that the process is in state j.

Corollary 5.11

Under the conditions of Theorem 5.8,

$$\lim_{t \to \infty} \{\text{Amount of time in state } j \text{ during } [0, t]\}/t = P_j$$

with probability 1, for all $j \geq 0$.

PROOF. If we let

$$f(x) = \begin{cases} 1 & \text{if } x = j \\ 0 & \text{if } x \neq j \end{cases}$$

then $\int_0^t f(X_s) \, ds$ represents the amount of time the process is in state j during $[0, t]$. But Theorem 5.10 asserts that $1/t \int_0^t f(X_s) \, ds \to P_j$ with probability 1, and hence the result is proven.

REMARK. If the process $\{X(t), t \geq 0\}$ becomes a regenerative process at some finite time T_0, then the results of this section remain true. That is, they are true if $\{X(T_0 + t), t \geq 0\}$ is a regenerative process where T_0 is finite with probability 1. Such a process is called a *delayed* or *general regenerative* process.

5.5. A Queueing Application

Suppose that customers arrive at a single-server service station in accordance with a nonlattice renewal process. Upon arrival, a customer is immediately served if the server is idle, and he waits in line if the server is busy. The service times of customers are assumed to be independent and identically distributed, and are also assumed independent of the arrival stream.

Let $X_1, X_2, \ldots$ denote the interarrival times between customers; and let $Y_1, Y_2, \ldots$ denote the service times of successive customers. We shall assume that

$$EY_1 < EX_1 < \infty \tag{15}$$

Suppose that the first customer arrives at time 0. If T_1 denotes the next time that a customer arrives and finds the server idle, then it is easy to see that T_1 is a regeneration point for the process $\{n(t), t \geq 0\}$, where $n(t)$ represents the number of customers in the system at t. Thus, if we call a period of time *busy* if the server is busy, and *idle* if the server is idle, then the process regenerates itself at the beginning of each busy period.

In order, to show that (15) implies that the expected length of a regeneration cycle is finite, we shall need the following lemma.

Lemma 5.12

Suppose $Z_1, Z_2, \ldots$ are independent and identically distributed with $EZ_1 > 0$.
If

$$N = \min\{n : Z_1 + \cdots + Z_n > 0\}$$

then

$$EN < \infty$$

PROOF. Let $S_n = \sum_{i=1}^{n} Z_i$. Since the Z_i's are independent and identically distributed, it follows that

$$P\{Z_1 \leq 0, Z_1 + Z_2 \leq 0, \ldots, Z_1 + \cdots + Z_n \leq 0\}$$
$$= P\{Z_n \leq 0, Z_n + Z_{n-1} \leq 0, \ldots, Z_n + \cdots + Z_1 \leq 0\}$$

Or equivalently,

$$P\{S_1 \leq 0, S_2 \leq 0, \ldots, S_n \leq 0\} = P\{S_n \leq S_{n-1}, S_n \leq S_{n-2}, \ldots, S_n \leq 0\}$$

However, the left side is just the probability that $N > n$, and hence by summing on n, we obtain

$$EN = \sum_{n=0}^{\infty} P\{S_n \leq S_{n-1}, S_n \leq S_{n-2}, \ldots, S_n \leq 0\} \tag{16}$$

Now, let us say that a renewal takes place at n if $S_n \leq S_{n-1}, S_n \leq S_{n-2}, \ldots,$ $S_n \leq 0$. (A little thought should convince us that the times between successive renewals are indeed independent and identically distributed.) However, by the strong law of large numbers, the total number of renewals is finite (since $S_n \leq 0$ only finitely often), and hence the expected number of renewals is finite (see Proposition 3.18). The result then follows, as the right side of (16) is just the expected number of renewals.

Let us now return to our queueing model. Recall that the first customer has just arrived at time 0. Now, suppose that $X_1 > Y_1$. Since X_1 represents the time until the next customer arrives, and Y_1 the service time of the initial customer, it would follow that the busy period ends at Y_1 and the regeneration cycle is of length X_1. In general, if we let

$$N = \min\{n : X_1 + \cdots + X_n > Y_1 + \cdots + Y_n\} \tag{17}$$

then the busy period will end at $Y_1 + \cdots + Y_N$, and the regeneration cycle will be of length

$$T_1 = X_1 + \cdots + X_N$$

However, by the previous lemma (since $EX_1 > EY_1$), we have that $EN < \infty$, and hence from Wald's equation we obtain

$$ET_1 = EX_1 \cdot EN < \infty \qquad (18)$$

Hence, from Theorem 5.8 it follows that

$$P_j = \lim_{t \to \infty} P\{n(t) = j\}$$

exists; and from Proposition 5.9 that

$$\int_0^t n(s)\, ds/t \to \frac{E \int_0^{T_1} n(s)\, ds}{ET_1} \qquad \text{as } t \to \infty \qquad \text{(with probability 1)}$$

and

$$E \int_0^t n(s)\, ds/t \to \frac{E \int_0^{T_1} n(s)\, ds}{ET_1} \qquad \text{as } t \to \infty$$

Let us denote by L, the limiting average (expected) number of customers in the system. That is,

$$L = \lim_{t \to \infty} \int_0^t n(s)\, ds/t$$

$$= \lim_{t \to \infty} E \int_0^t n(s)\, ds/t \qquad (19)^\dagger$$

$$= \frac{E \int_0^{T_1} n(s)\, ds}{ET_1}$$

Now, let W_n, $n = 1, 2, \ldots$ denote the amount of time that the nth customer spends in the system. W_n is often called the *waiting time* of the nth customer. Since N, as given by (17), represents the total number of customers served in a regeneration cycle, it follows that $W_1 + \cdots + W_N$ has the same distribution as $W_{N+1} + \cdots + W_{N+N_2}$, where N_2 represents the number of customers served in the second cycle. Hence, by regarding the total waiting time of customers of the nth cycle as the reward for that cycle, we have by Proposition 3.16 that

$$\sum_{i=1}^n \frac{W_i}{n} \to \frac{E \sum_{i=1}^N W_i}{EN} \qquad \text{as } n \to \infty \qquad \text{(with probability 1)}$$

and

$$E \sum_{i=1}^n \frac{W_i}{n} \to \frac{E \sum_{i=1}^N W_i}{EN} \qquad \text{as } n \to \infty$$

† Note that from Theorem 5.10 we also have that $L = \sum_{j=0}^{\infty} jP_j$.

Thus, letting W represent the average (expected) wait of a customer, we have

$$
\begin{aligned}
W &= \lim_{n \to \infty} \sum_{i=1}^{n} \frac{W_i}{n} \\
&= \lim_{n \to \infty} E \sum_{i=1}^{n} \frac{W_i}{n} \\
&= \frac{E \sum_{i=1}^{N} W_i}{EN}
\end{aligned}
\tag{20}
$$

The important relationship between L, the time average number of customers in the system, and W, the average waiting time of a customer is given by the following theorem.

Theorem 5.13

Let $\lambda = 1/EX_1$, then

$$L = \lambda W$$

PROOF.   From (19) and (18) we have that

$$
\begin{aligned}
L &= \frac{E \int_0^{T_1} n(s)\, ds}{ET_1} \\
&= \frac{E \int_0^{T_1} n(s)\, ds}{EN} \cdot \frac{1}{EX_1}
\end{aligned}
\tag{21}
$$

Now, from (20),

$$
W = \frac{E \sum_{i=1}^{N} W_i}{EN}
\tag{22}
$$

and the result follows from (21) and (22), since

$$
\sum_{i=1}^{N} W_i = \int_0^{T_1} n(s)\, ds
$$

This last equality follows since $\sum_{i=1}^{N} W_i$ and $\int_0^{T_1} n(s)\, ds$ both represent the total number of waiting hours in a cycle (see Figure 5.1). [Both $W_1 + \cdots + W_N$ and $\int_0^{T_1} n(t)\, dt$ represent the area of the enclosed curve.]

REMARKS.   The proof of Theorem 5.13 does not depend on the particular queueing model we have assumed. The proof goes through without change for any queueing system that contains regeneration points and is such that the mean length of a cycle is finite. For example, if in our model we suppose that

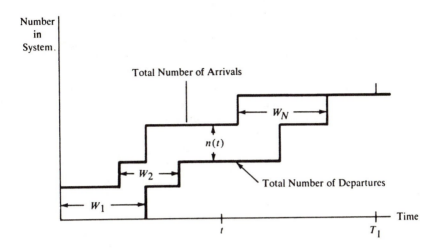

Figure 5.1. Typical Realization of a Regeneration Cycle

there are c available servers, then it can be shown that the mean cycle length is finite if and only if

$$EY_1 < cEX_1 \qquad (23)$$

Hence, if (23) is satisfied, then Theorem 5.13 remains valid for this many server queueing model.

Furthermore, it is not difficult to show that $\lim_{n\to\infty} P\{W_n < a\}$ exists, and that W is just the mean of this limiting distribution (see Problem 4). Hence, $L = \lambda W$ may also be interpreted as asserting that the mean of the limiting (as $t \to \infty$) distribution of the number of customers in the system is equal to λ times the mean of the limiting (as $n \to \infty$) distribution of the waiting time per customer.

5.6. Back to Markov Renewal Processes—Limiting Probabilities

As a straightforward application of Theorem 5.8 (and the final remark of that section), we have the following proposition.

Proposition 5.14

If $i \in C_j$, G_{jj} is not lattice, and $\mu_{jj} < \infty$, then

$$P_j = \lim_{t\to\infty} P\{Z(t) = j \mid Z(0) = i\} = \frac{\mu_j}{\mu_{jj}} \qquad (24)$$

Using (24), it is not difficult to show that positive (null) recurrence is a class property (see Problem 11).

Let us suppose for the remainder of this section that the Markov renewal process is irreducible, positive recurrent and not lattice. Suppose further, that the embedded Markov chain $\{J_n, n \geq 0\}$ is also positive recurrent and aperiodic, and let $\{\pi_j, j = 0, 1, 2, ..\}$ denote the limiting probabilities for this Markov chain.

Since π_j equals the (long-run) proportion of transitions which are into state j, and since μ_j is the mean time spent in j per transition, it seems intuitive that the limiting probabilities should be proportional to $\pi_j \mu_j$. That is,

$$P_j = \frac{\pi_j \mu_j}{\sum_{i=0}^{\infty} \pi_i \mu_i} \qquad j = 0, 1, 2, \ldots \tag{25}$$

In order to prove (25), we shall need the following proposition which is useful in its own right.

Proposition 5.15

For all i, j

$$\mu_{ij} = \mu_i + \sum_{k \neq j} P_{ik} \mu_{kj} \tag{26}$$

PROOF.   Let T_{ij} denote the time of the first transition into j. Then

$$\mu_{ij} = \sum_{k=0}^{\infty} E[T_{ij} \mid J_1 = k, J_0 = i] P_{ik}$$

$$= \sum_{k \neq j} P_{ik}[\mu_{kj} + \eta_{ik}] + P_{ij} \eta_{ij}$$

$$= \sum_{k=0}^{\infty} P_{ik} \eta_{ik} + \sum_{k \neq j} P_{ik} \mu_{kj}$$

$$= \mu_i + \sum_{k \neq j} P_{ik} \mu_{kj}$$

We are now ready to prove the important

Theorem 5.16

For all states j,

$$\mu_{jj} \pi_j = \sum_{i=0}^{\infty} \pi_i \mu_i$$

and

$$P_j = \frac{\pi_j \mu_j}{\sum_{i=0}^{\infty} \pi_i \mu_i}$$

PROOF.   Multiplying both sides of (26) by π_i and summing on i yields

$$\sum_{i=0}^{\infty} \pi_i \mu_{ij} = \sum_{i=0}^{\infty} \pi_i \mu_i + \sum_{i=0}^{\infty} \pi_i \sum_{k \neq j} P_{ik} \mu_{kj}$$

$$= \sum_{i=0}^{\infty} \pi_i \mu_i + \sum_{k \neq j} \mu_{kj} \sum_{i=0}^{\infty} \pi_i P_{ik}$$

$$= \sum_{i=0}^{\infty} \pi_i \mu_i + \sum_{k \neq j} \pi_k \mu_{kj}$$

or

$$\pi_j \mu_{jj} = \sum_{i=0}^{\infty} \pi_i \mu_i \tag{27}$$

where we have used the fact that $\pi_k = \sum_{i=0}^{\infty} \pi_i P_{ik}$ (see 4.9). From (27) it follows that

$$\frac{\pi_j \mu_j}{\sum_{i=0}^{\infty} \pi_i \mu_i} = \frac{\mu_j}{\mu_{jj}}$$

and the result follows from Proposition 5.14.

REMARK.   Theorem 5.16 is extremely important, for it reduces the problem of calculating limiting probabilities of the Markov renewal process to one of calculating the stationary probabilities of the embedded Markov chain. It also shows that the limiting probabilities only depend on $Q_{ij}(\cdot)$ thru P_{ij} and μ_j.

EXAMPLE.   The (M/M/1) Queue.   Consider a single-server queueing system in which customers arrive at a Poisson rate λ, and service times are independent exponential random variables having mean $1/\mu$. Letting $n(t)$ denote the number of customers in the system at time t, then because of the lack of memory of the exponential it follows that the process $\{n(t), t \geq 0\}$ is semi-Markov with

$$P_{01} = 1 \qquad \mu_0 = \frac{1}{\lambda}$$

$$P_{i,i-1} = \frac{\mu}{\lambda + \mu} \qquad \mu_i = \frac{1}{\lambda + \mu} \qquad i = 1, 2, \dots$$

$$P_{i,i+1} = \frac{\lambda}{\lambda + \mu} \qquad\qquad\qquad i = 1, 2, \dots$$

Suppose that $\lambda < \mu$. The limiting probabilities for the embedded chain are given by

$$\pi_0 = \pi_1 \frac{\mu}{\lambda + \mu}$$

$$\pi_1 = \pi_0 + \pi_2 \frac{\mu}{\lambda + \mu}$$

$$\pi_i = \pi_{i-1} \frac{\lambda}{\lambda + \mu} + \pi_{i+1} \frac{\mu}{\lambda + \mu} \qquad i = 2, 3, \ldots$$

These equations are easily solved in terms to π_0 to yield

$$\pi_0 = \pi_0$$

$$\pi_i = \pi_0 (\lambda/\mu)^{i-1} \frac{\lambda + \mu}{\mu} \qquad i = 1, 2, \ldots$$

Hence,

$$\pi_0 \mu_0 = \pi_0 / \lambda$$

$$\pi_i \mu_i = \frac{\pi_0}{\lambda} (\lambda/\mu)^i \qquad i = 1, 2, \ldots$$

And so by Theorem 5.16,

$$P_i = (\lambda/\mu)^i P_0$$

and since the probabilities must sum to unity, we obtain

$$P_i = (\lambda/\mu)^i (1 - \lambda/\mu) \quad i = 0, 1, 2, \ldots$$

EXAMPLE. *The M/M/1 Queue with Batch Service.* Consider the previous model, but suppose now that if there are two or more customers waiting when the server is free, then he serves the next two together. The service time is assumed to be exponential with mean $1/\mu$, regardless of the number being served.

The above is a semi-Markov process with states $0, 1, 2, \bar{2}, 3, \bar{3}, 4, \bar{4}, \ldots$, where state n means "n customers in the system and one in service," and state $\bar{n}$ means "n customers in the system and two in service." The transition probabilities are given by

$$P_{01} = 1$$

$$P_{i, i+1} = \frac{\lambda}{\lambda + \mu}, \qquad i \geq 1$$

$$P_{i,\,i-1} = \frac{\mu}{\lambda+\mu}, \qquad i = 1, 2$$

$$P_{i,\,\overline{i-1}} = \frac{\mu}{\lambda+\mu}, \qquad i > 2$$

$$P_{i,\,\overline{i+1}} = \frac{\lambda}{\lambda+\mu}, \qquad i \geq 2$$

$$P_{i,\,i-2} = \frac{\mu}{\lambda+\mu}, \qquad i = 2, 3$$

$$P_{i,\,\overline{i-2}} = \frac{\mu}{\lambda+\mu}, \qquad i > 3$$

Hence, the limiting probabilities of the embedded chain are determined by

$$\pi_0 = \pi_1 \frac{\mu}{\lambda+\mu} + \pi_{\overline{2}} \frac{\mu}{\lambda+\mu}$$

$$\pi_1 = \pi_0 + \pi_2 \frac{\mu}{\lambda+\mu} + \pi_3 \frac{\mu}{\lambda+\mu}$$

$$\pi_i = \pi_{i-1} \frac{\lambda}{\lambda+\mu}, \qquad\qquad i \geq 2$$

$$\pi_{\overline{2}} = \pi_3 \frac{\mu}{\lambda+\mu} + \pi_{\overline{4}} \frac{\mu}{\lambda+\mu}$$

$$\pi_i = \pi_{i+1} \frac{\mu}{\lambda+\mu} + \pi_{\overline{i+2}} \frac{\mu}{\lambda+\mu} + \pi_{\overline{i-1}} \frac{\lambda}{\lambda+\mu}, \qquad i > 2$$

$$\sum_{i=0}^{\infty} \pi_i + \sum_{i=2}^{\infty} \pi_{\overline{i}} = 1$$

These equations may be solved and the limiting probabilities would then be given by

$$P_x = \frac{\pi_x \mu_x}{\sum_x \pi_x \mu_x}, \qquad x = 0, 1, 2, \overline{2}', 3, \overline{3}', \ldots$$

where

$$\mu_x = \begin{cases} \dfrac{1}{\lambda} & \text{for } x = 0 \\[2mm] \dfrac{1}{\lambda+\mu} & \text{for } x \neq 0 \end{cases}$$

5.7. Limiting Distribution of the Markov Renewal Process

The problem of obtaining the limiting distribution of a Markov renewal process is not completely solved by deriving the limits of the $P_{ij}(t)$. One must instead consider the problem of finding the limit, as $t \to \infty$, of the probability of being in state j at time t, of making the next transition sometime before $t + x$ and of this next transition being into state k.

As a first step, we shall consider this probability when the initial state is j. That is,

$$P\{Z(t) = j, J_{N(t)+1} = k, Y(t) \le x \,|\, J_0 = j\} \tag{28}$$

where $Y(t)$ denotes the time from t until the next transition (that is, the excess time at t).

Suppressing the dependence on j, k, x, let $R(t)$ denote the probability (28). If j is recurrent, then by conditioning on T, the time of the first transition into j, we obtain

$$R(t) = \int_0^t R(t - s) \, dG_{jj}(s) + h(t) \tag{29}$$

where

$$h(t) = \int_t^\infty P\{Z(t) = j, J_{N(t)+1} = k, Y(t) \le x \,|\, J_0 = j, T = s\} \, dG_{jj}(s)$$

$$= \int_t^\infty P\{X_1 > t, J_1 = k, X_1 \le t + x \,|\, J_0 = j, T = s\} \, dG_{jj}(s)$$

This last equation follows, since if $T = s > t$, then the only way that $Z(t)$ can be equal to j is for the process not to have made any transitions by time t. (Recall that X_1 is the time until the first transition.) However,

$$P\{t < X_1 \le t + x, J_1 = k \,|\, J_0 = j\}$$

$$= \int_0^\infty P\{t < X_1 \le t + x, J_1 = k \,|\, J_0 = j, T = s\} \, dG_{jj}(s) \tag{30}$$

$$= h(t)$$

where the integral in (30) only goes from t to infinity, since $X_1 \le T$.

By Proposition 3.4, the solution of the renewal type equation (29) is given by

$$R(t) = h(t) + \int_0^t h(t - x) \, dm(x)$$

where

$$m(x) = \sum_{n=1}^\infty (G_{jj})_n(x)$$

and thus by the key renewal theorem (assuming that G_{jj} is not lattice),

$$\lim_{t \to \infty} R(t) = \int_0^\infty \frac{h(t)\,dt}{\mu_{jj}}$$

$$= \frac{1}{\mu_{jj}} \int_0^\infty P\{t < X_1 \le t + x, J_1 = k \mid J_0 = j\}\,dt$$

$$= \frac{1}{\mu_{jj}} \int_0^\infty P_{jk}[F_{jk}(t + x) - F_{jk}(t)]\,dt \tag{31}$$

$$= \frac{P_{jk}}{\mu_{jj}} \int_0^\infty \{1 - F_{jk}(t) - [1 - F_{jk}(t + x)]\}\,dt$$

$$= \frac{P_{jk}}{\mu_{jj}} \int_0^x [1 - F_{jk}(t)]\,dt$$

If the process is irreducible, then it is easy to prove (using the above) that (31) remains true even when the initial state is not j. Also, note that (31) is also valid when j is transient, since both sides would equal 0. Summarizing, we have proven the following theorem.

Theorem 5.17

If the process is irreducible and not lattice, then

$$P\{Z(t) = j, J_{N(t)+1} = k, Y(t) \le x \mid J_0 = i\} \to \frac{P_{jk}}{\mu_{jj}} \int_0^x [1 - F_{jk}(t)]\,dt$$

Notice that when there is only one state (that is, when we have a renewal process), Theorem 5.17 yields the limiting distribution of excess.

Also, in analogy with results proven for renewal processes (Proposition 3.15) and for Markov chains (Theorem 4.9) it may be shown that if

   (i) the Markov renewal process is positive recurrent and irreducible, and if
   (ii) the joint distribution of the initial state, the amount of time spent in that state before transition, and the next state are given by (31),

then the process $\{Z(t), J_{N(t)+1}, Y(t), t \ge 0\}$ is stationary with a joint distribution (for each t) given by (31).

5.8. Continuous Time Markov Chains

Consider a semi-Markov process $\{Z(t),\ t \geq 0\}$ for which

$$P_{ii} = 0 \qquad i = 0, 1, 2, \ldots$$

and

$$F_{ij}(t) = 1 - e^{-\lambda_i t} \qquad i = 0, 1, 2, \ldots$$

Because of the exponential character of $F_{ij}(t)$ (which, it should be noted, is independent of j), it is easily verified that $\{Z(t),\ t \geq 0\}$ is a Markov process. That is,

$$P\{Z(t) = j \mid Z(s),\ s \leq t_0\} = P\{Z(t) = j \mid Z(t_0)\}$$

for all j, $t_0 \leq t$. Such a Markov process is called a *continuous time Markov chain*.

By exploiting the Markovian property, we are able to derive two sets of differential equations for $P_{ij}(t)$, which may sometimes be explicitly solved. However, before doing so we shall first need the following lemmas.

Lemma 5.18

(a) $\displaystyle \lim_{t \to 0} \frac{1 - P_{ii}(t)}{t} = \lambda_i$

(b) $\displaystyle \lim_{t \to 0} \frac{P_{ij}(t)}{t} = \lambda_i P_{ij}$

Lemma 5.19

For all s, t

$$P_{ij}(t + s) = \sum_{k=0}^{\infty} P_{ik}(t) P_{kj}(s)$$

Lemma 5.18 is a direct consequence of the fact that the probability of two or more transitions in time t is $o(t)$, while Lemma 5.19 follows directly from the Markovian property.

From Lemma 5.19 we obtain,

$$P_{ij}(t + h) - P_{ij}(t) = \sum_{k \neq i} P_{ik}(h) P_{kj}(t) - [1 - P_{ii}(h)] P_{ij}(t)$$

Hence,

$$\lim_{h \to 0} \frac{P_{ij}(t + h) - P_{ij}(t)}{h} = \lim_{h \to 0} \sum_{k \neq i} \frac{P_{ik}(h)}{h} P_{kj}(t) - \lambda_i P_{ij}(t) \qquad (32)$$

Now, for any fixed N,

$$\liminf_{h \to 0} \sum_{k \neq i} \frac{P_{ik}(h)}{h} P_{kj}(t) \geq \liminf_{h \to 0} \sum_{\substack{k \neq i \\ k=0}}^{N} \frac{P_{ik}(h)}{h} P_{kj}(t)$$

$$= \lambda_i \sum_{\substack{k \neq i \\ k=0}}^{N} P_{ik} P_{kj}(t)$$

implying that

$$\liminf_{h \to 0} \sum_{k \neq i} \frac{P_{ik}(h)}{h} P_{kj}(t) \geq \lambda_i \sum_{k \neq i} P_{ik} P_{kj}(t) \tag{33}$$

On the other hand, for N large,

$$\limsup_{h \to 0} \sum_{k \neq i} \frac{P_{ik}(h)}{h} P_{kj}(t) \leq \limsup_{h \to 0} \left[\sum_{\substack{k \neq i \\ k=0}}^{N} \frac{P_{ik}(h)}{h} P_{kj}(t) + \sum_{k=N+1}^{\infty} \frac{P_{ik}(h)}{h} \right]$$

$$= \limsup_{h \to 0} \left[\sum_{\substack{k \neq i \\ k=0}}^{N} \frac{P_{ik}(h)}{h} P_{kj}(t) + \frac{1}{h} \left(1 - \sum_{k=0}^{N} P_{ik}(h) \right) \right]$$

$$= \lambda_i \sum_{\substack{k \neq i \\ k=0}}^{N} P_{ik} P_{kj}(t) + \lambda_i - \lambda_i \sum_{k=0}^{N} P_{ik}$$

and by letting $N \to \infty$, we obtain

$$\limsup_{h \to 0} \sum_{k \neq i} \frac{P_{ik}(h)}{h} P_{kj}(t) \leq \lambda_i \sum_{k \neq i} P_{ik} P_{kj}(t)$$

which, along with (33), yields,

$$\lim \sum_{k \neq i} \frac{P_{ik}(h)}{h} P_{kj}(t) = \lambda_i \sum_{k \neq i} P_{ik} P_{kj}(t) \tag{34}$$

Thus, combining (34) with (32), we obtain

Theorem 5.20

For all $i, j, t \geq 0$,

$$P'_{ij}(t) = \lambda_i \sum_{k \neq i} P_{ik} P_{kj}(t) - \lambda_i P_{ij}(t) \tag{35}$$

The set of equations (35) are known as *Kolmogorov's backward equations.*
We may derive another set of equations, known as *Kolmogorov's forward equations*, in the following manner.

$$\frac{P_{ij}(t+h) - P_{ij}(t)}{h} = \sum_{k \neq j} P_{ik}(t) \frac{P_{kj}(h)}{h} - P_{ij}(t) \left(\frac{1 - P_{jj}(h)}{h} \right)$$

And assuming that we can take the limit inside the summation sign, we obtain (by letting $h \to 0$),

$$P'_{ij}(t) = \sum_{k \neq j} \lambda_k P_{kj} P_{ik}(t) - \lambda_j P_{ij}(t) \tag{36}$$

Unfortunately, we cannot always justify the interchange of limit and summation, and thus (36) is not always valid. Two sufficient conditions for its validity are (i) that the state space is finite; or (ii) if

$$\lim_{t \to 0} \frac{P_{ij}(t)}{t} \to \lambda_i P_{ij}$$

uniformly in i for $i \neq j$.

5.9. Birth and Death Processes

A continuous time Markov chain $\{Z(t), t \geq 0\}$ with state space $0, 1, 2, \ldots$, for which $P_{ij} = 0$ whenever $|i - j| \geq 2$, is called a *birth and death* process. In other words, a birth and death process is a continuous time Markov chain which changes only through transitions from a state to its immediate neighbor.

For a birth and death process, it is convenient to introduce the quantities v_j and θ_j defined as follows:

$$\theta_0 = 0$$
$$\theta_j = \lambda_j P_{j, j-1} \qquad \text{for } j \geq 1$$
$$v_j = \lambda_j P_{j, j+1} \qquad \text{for } j \geq 0$$

From lemma 5.18 we have that

$$\lim_{t \to 0} \frac{P_{j, j-1}(t)}{t} = \theta_j, \qquad j \geq 0$$

$$\lim_{t \to 0} \frac{P_{j, j+1}(t)}{t} = v_j, \qquad j \geq 0$$

$$\lim_{t \to 0} \frac{1 - P_{jj}(t)}{t} = \theta_j + v_j, \qquad j \geq 0$$

If we think of $Z(t)$ as the size of a population at time t, then transitions from j to $j + 1$ are called *births*, and those from j to $j - 1$ are called *deaths*. A valuable way of interpreting the birth and death process is as follows: When the population size is j, then the time until the next birth has an exponential distribution with mean $1/v_j$, and the time until the next death is exponential with mean $1/\theta_j$.

Examples of Birth and Death Processes

1. *The* M/M/1 *Queue.*
If

$$v_j = v, \qquad j \geq 0$$
$$\theta_j = \theta, \qquad j \geq 1$$

then we obtain the single-server Poisson queue with exponential service times.

2. *The* M/M/1 *Queue with Balking*
If

$$v_j = \alpha_j v, \qquad j \geq 0$$
$$\theta_j = \theta, \qquad j \geq 1$$

where $0 \leq \alpha_j \leq 1$, then this birth and death process is an $M/M/1$ queueing system with balking. That is, a customer who arrives when there are j persons presently in the system will enter the queueing system with probability α_j; the probability that he will not enter the system (i.e., that he will balk) is $1 - \alpha_j$.

A special case of this model occurs when

$$\alpha_j = \begin{cases} 1 & j \leq N \\ 0 & j > N \end{cases}$$

This is the $M/M/1$ queueing system with finite capacity N. That is, the queueing system can hold at most N customers, and all customers who arrive when the system is full are lost.

3. *The* M/M/c *Queue*
If

$$v_j = v, \qquad j \geq 0$$
$$\theta_j = \begin{cases} j\theta & j \leq c \\ c\theta & j > c \end{cases}$$

then we obtain the c server, Poisson arrival, exponential service time queueing system.

4. *Linear Growth with Immigration*
If

$$v_j = jv + a, \qquad j \geq 0$$
$$\theta_j = j\theta, \qquad j \geq 0$$

then we obtain what is called a *linear growth process*. Such processes occur naturally in the study of biological reproduction and population growth. If the current population size is j, then each individual gives birth at an exponential rate v; furthermore, there is an exponential rate of increase a of the population due to an external source such as immigration. Hence, the total birth rate when there are j in the system is $jv + a$. Deaths are assumed to occur at an exponential rate θ for each member of the population, and hence $\theta_j = j\theta$.

We shall now determine the limiting probabilities P_j of the birth and death process. The equations determining the limiting probabilities π_j of the embedded Markov chain are by Proposition 4.9.

$$\pi_0 = \pi_1 \frac{\theta_1}{\theta_1 + v_1}$$

$$\pi_j = \pi_{j+1} \frac{\theta_{j+1}}{\theta_{j+1} + v_{j+1}} + \pi_{j-1} \frac{v_{j-1}}{\theta_{j-1} + v_{j-1}}, \qquad j \geq 1$$

Hence, since

$$\frac{\pi_j}{\theta_j + v_j} = cP_j$$

(assuming these limiting probabilities exist), we obtain

$$v_0 P_0 = \theta_1 P_1$$
$$(v_j + \theta_j)P_j = \theta_{j+1}P_{j+1} + v_{j-1}P_{j-1}, \qquad j \geq 1$$

Solving in terms of P_0 yields, by induction,

$$P_j = b_j P_0, \qquad j \geq 1$$

where

$$b_0 = 1, \ b_j = \prod_{i=0}^{j-1} \frac{v_i}{\theta_{i+1}} \tag{37}$$

Thus,

$$P_0 \sum_{j=0}^{\infty} b_j = 1$$

hence,

$$P_j = \frac{b_j}{\sum_{n=0}^{\infty} b_n} \tag{38}$$

Therefore, clearly a necessary condition for these limiting probabilities to exist is that $\sum_{n=0}^{\infty} b_n < \infty$. It may be shown (see, for instance, Chung [3]) that this condition is also sufficient; hence if $\sum_{n=0}^{\infty} b_n < \infty$, then Equations (37) and (38) yield the limiting probabilities.

A birth and death process is said to be a *pure birth* process if $\theta_j = 0$ for all j, (that is, if death is impossible).

The Kolmogorov forward equations (36) for a pure birth process become

$$P'_{ii}(t) = -v_i P_{ii}(t)$$
$$P'_{ij}(t) = v_{j-1}P_{i,\,j-1}(t) - v_j P_{ij}(t) \tag{39}$$

Proposition 5.21

The solution of Equation (39) is

$$P_{ii}(t) = e^{-v_i t}, \qquad\qquad\qquad i \geq 0$$

$$P_{ij}(t) = v_{j-1}e^{-v_j t} \int_0^t e^{v_j s}P_{i,\,j-1}(s)\,ds, \qquad j \geq i+1$$

PROOF. The fact that $P_{ii}(t) = e^{-v_i t}$ follows directly from Equation (39). To prove the corresponding result for $P_{ij}(t)$, we note by (39) that

$$e^{v_j t}[P'_{ij}(t) + v_j P_{ij}(t)] = e^{v_j t}v_{j-1}P_{i,\,j-1}(t)$$

or

$$\frac{d}{dt}[e^{v_j t}P_{ij}(t)] = v_{j-1}e^{v_j t}P_{i,\,j-1}(t)$$

Hence, since $P_{ij}(0) = 0$, we obtain the desired result.

Examples of Pure Birth Processes

1. *The Poisson Process*
The pure birth process having $v_j = v$ for all j is, of course, the Poisson process.
2. *The Birth Process with Linear Birthrate*
Consider a population whose members can give birth to new members but cannot die. If each member acts independently of the others and takes an exponential amount of time to give birth, then we have a pure birth process with

$$v_j = jv, \qquad j \geq 0$$

Using Proposition 5.21 and mathematical induction, it is not difficult to show that for $j \geq i \geq 1$

$$P_{ij}(t) = \binom{j-1}{j-i}(e^{-vt})^i(1 - e^{-vt})^{j-i}$$

This pure birth process is known as a *Yule process*.

Problems

1. Show that J_0 and $\{\bar{N}(t), t \geq 0\}$ determine $\{Z(t), t \geq 0\}$.

2. Transitions may be thought of as taking place in two stages in the following manner. The time until a transition from i occurs is chosen according to some distribution $R_i(t)$, and given that the time selected is t, the next state is chosen according to P^t_{ij}.

$$R_i(t) = ?, \quad P^t_{ij} = ?$$

3. Give an example of a positive (null) recurrent Markov renewal process whose embedded chain in null (positive) recurrent.

4. Prove Proposition 5.6.

5. Under the conditions of Theorem 5.8, show that P_j is the limiting proportion (as $n \to \infty$) of time spent in state j during the first n cycles.

6. Consider an (s, S) inventory ordering policy. Find the limiting distribution of X_n, the inventory level at time n, if there is a lag of one time period between the time at which the order is placed and the time it is received. Assume an exponential demand distribution.

7. State and prove the equivalent of Theorem 5.8 for a regenerative process whose state space is the real line.

8. Show that $\{W_n, n = 1, 2, \ldots\}$ as defined in Section 5 is a discrete time regeneration process with an uncountable state space. Find $W_\infty(a) = \lim_{n \to \infty} P\{W_n < a\}$, and show that W equals the mean of this limiting distribution.

9. For the queueing example of Section 5, find the mean length of a busy period, and the mean number of customers served during a busy period when the arrival stream is Poisson.

10. For the queueing example of Section 5, assume that the arrival stream is a Poisson process and find (i) P_0, and (ii) the proportion of customers who don't have to wait in queue before receiving service.

11. Show that positive recurrence is a class property.

12. Show that

$$\mu_{jj} = \mu_j + \sum_{i \neq j} \mu_i E[_jN_i]$$

where $_jN_i$ denotes the number of visits to i prior to the first return to j.

13. For a positive recurrent irreducible Markov renewal process, show that

$$E[_jN_i] = \frac{\mu_{jj}}{\mu_{ii}}$$

14. Determine the limiting probabilities for the $M/M/c$ queueing system.

15. *A Markov renewal reward process.* Consider a positive recurrent irreducible Markov renewal process in which a reward, distributed according to F_i, is earned each time a transition into i occurs. Suppose successive rewards are independent and let $Y(t)$ denote the total reward earned by t.

$$\frac{Y(t)}{t} \to ?, \quad E\frac{Y(t)}{t} \to ? \quad \text{(Prove)}$$

16. Prove the assertion made in the last paragraph of Section 7.

17. Prove Lemma 5.18.

18. For a continuous time Markov chain, show that

$$m_{jj}(t) = \frac{t}{\mu_{jj}} - 1 + \frac{1}{\mu_{jj}} \sum_{k=0}^{\infty} P_{jk}(t)\mu_{kj}$$

19. Show that the pure birth process is regular if and only if $\sum_{i=1}^{\infty} 1/v_i = \infty$.

20. Consider two machines, both of which have an exponential lifetime with mean $1/\lambda$. The time to repair for each machine has the same exponential distribution with mean $1/\mu$. Given that at time 0 both machines are in working condition, find the probability that they will both be operative at time t. Assume independence.

21. Consider a continuous time Markov chain with state space $\{0, 1\}$. Find $P_{ij}(t)$ by solving the Kolmogorov equations. Note that this is just an alternating renewal process for which the amount of time spent in each state is exponential.

22. For the linear growth process with immigration, show, by using the forward equations, that $M(t) \equiv E[z(t)]$ satisfies

$$M'(t) = a + (v - \theta)M(t)$$

and solve this equation to obtain

$$M(t) = \begin{cases} at + z(0) & \text{if } v = \theta \\ \dfrac{a}{v - \theta} \{e^{(v-\theta)t} - 1\} + z(0)e^{(v-\theta)t} & \text{if } v \neq \theta \end{cases}$$

Discuss the limiting behaviour (as $t \to \infty$) of $M(t)$.

23. For a birth and death process, show that the expected length of time for reaching state $n + 1$ starting from state 0 is

$$\sum_{j=0}^{n} P_j \sum_{i=0}^{j} 1/\lambda_i P_i$$

where P_j, $j \geq 0$ are the limiting probabilities given by (38).

24. Consider a Yule process $\{z(t), t \geq 0\}$ with birthrate v and initial population of size 1. Find the distribution of the number of members of the population at time t of age less than or equal to x.

References

A basic reference for the first three sections consists of the papers by R. Pyke [5] and [6]. We have basically followed his approach, though our proofs are not always alike.

The idea of a regenerative process originated with W. Feller in his work on recurrent events and culminated with the work of W. Smith (see [7] and [8]). Theorem 5.8 is due to Smith. Theorem 5.10 may be found in a recent paper by M. Brown and S. Ross [2]. The argument that the mean regeneration time is finite for the queueing model in Section 5 is also due to Smith [8]. Theorem 5.13 is due to Little. However, the proof given is an adaptation of one due to W. Jewell [4]. Theorem 5.16 is due to R. Barlow [1]. The results in Section 7 are due to Pyke [6]. For further results on continuous time Markov chains, the interested reader is referred to Chung [3].

Note also that in the statement of Theorem 5.8, we require that T_1 have an absolutely continuous component. This condition, which is stronger than the condition that T_1 be non-lattice, is necessary to ensure that $q_j(t)$ is directly Riemann integrable. A counterexample showing that the theorem need not hold when T_1 is non-lattice is the following:

$$\text{Let } X(t) = \begin{bmatrix} 1 & t \text{ rational} \\ 0 & t \text{ irrational} \end{bmatrix}$$

and let T_1 be a random variable having a non-lattice distribution concentrated on the rationals. Then T_1 is a regeneration point for the $X(t)$ process, but clearly $P[X(t) = 0]$ has no limit. This example is due to Mark Brown.

[1] BARLOW, R. "Applications of Semi-Markov Processes to Counter Problems," *Studies in Applied Probability and Management Science*, Chap. 3, (K. J. Arrow, S. Karlin, and H. Scarf, editors), Stanford University Press, Stanford, California, (1962).

[2] BROWN, M. and S. ROSS. "Renewal Reward Processes," unpublished manuscript.

[3] CHUNG, K. L. "*Markov Chains with Stationary Transition Probabilities*," Springer-Verlag, Berlin, (1960).

[4] JEWELL, W. S. "A Simple Proof of $L = \lambda W$," *Operations Research*, 15, pp. 1109–1116, (1967).

[5] PYKE, R. "Markov Renewal Processes: Definitions and Preliminary Properties," *Annals of Mathematical Statistics*, 32, pp. 1231–1242, (1961).

[6] PYKE, R. "Markov Renewal Processes with Finitely Many States," *Annals of Mathematical Statistics*, 32, pp. 1243–1259, (1961).

[7] SMITH, W. L. "Regenerative Stochastic Processes," *Proceedings Royal Society*, Series A, 232, pp. 6–31, (1955).

[8] SMITH, W. L. "Renewal Theory and Its Ramifications," *Journal of the Royal Statistical Society*, Series B, 20, pp. 243–302, (1958).

6

MARKOV DECISION PROCESSES

6.1. Introduction

In this chaper we shall consider a process which is observed at time points $t = 0, 1, 2, \ldots$ to be in one of a number of possible states. The set of possible states is assumed to be countable and will be labeled by the nonnegative integers $0, 1, 2 \ldots$. After observing the state of the process, an action must be chosen, and we let A, assumed finite, denote the set of all possible actions.

If the process is in state i at time t and action a is chosen, then two things occur:

(i) We incur a cost $C(i, a)$.

(ii) The next state of the system is chosen according to the transition probabilities $P_{ij}(a)$.

If we let X_t denote the state of the process at time t, and a_t the action chosen at t, then assumption (ii) is equivalent to stating that

$$P\{X_{t+1} = j \mid X_0, a_0, X_1, a_1, \ldots, X_t = i, a_t = a\} = P_{ij}(a)$$

Thus, both the costs and the transition probabilities are functions only of the last state and the subsequent action. Furthermore, we shall suppose that the costs are bounded and we let M be such that $|C(i, a)| < M$ for all i, a.

In order to choose actions, we must follow some policy. We shall place no restrictions on the class of allowable policies and we therefore define a *policy* to be any rule for choosing actions. Thus, the action chosen by a policy may, for instance, depend on the history of the process up to that point, or it may be randomized in the sense that it chooses action a with some probability $P_a, a \in A$.

An important subclass of the class of all policies is the class of stationary policies, where a policy is said to be *stationary* if it is nonrandomized and the action it chooses at time t only depends on the state of the process at time t. Or, in other words, a stationary policy is a function f mapping the state space

119

into the action space. Also, it easily follows that if a stationary policy f is employed, then the sequence of states $\{X_t, t = 0, 1, 2, \ldots\}$ forms a Markov chain with transition probabilities $P_{ij} = P_{ij}[f(i)]$; and it is for this reason that the process is called a Markov Decision Process.

This formulation of a Markov decision process is quite broad, as it includes a great many optimization models as special cases. In this chapter, we shall attempt to develop a theory which will enable us to find policies that are, in some sense, optimal. However, in order to do this we need to first decide upon appropriate optimality criteria.

In Section 6.2, we shall consider the criterion of the total expected discounted cost. We shall develop a theory concerning the existence of an optimal policy and then present methods for obtaining the optimal policy. The results of this section will then be applied in Section 6.3 to such diverse areas as (a) determining the optimal time to replace a machine, (b) maintaining the quality of manufactured goods and (c) determining the optimal policy for selling a house.

In Section 6.4, we shall consider the criterion of the total expected cost. The theory developed in this section will then be applied to (a) the theory of optimal stopping and (b) the statistical theory of sequential analysis.

In Sections 6.6 and 6.7, we shall consider the average cost criterion. It will first be shown, by means of counterexamples, that certain intuitive results do not hold when this criterion is employed. For instance, an example will be presented for which an optimal policy exists but no stationary policy is optimal.

In the final section of this chapter, some computational approaches will be suggested when the state space is finite.

6.2. Expected Discounted Cost

Functional Equation

This criterion assumes a discount factor $\alpha \in (0, 1)$, and attempts to minimize the expected total discounted cost.

For any policy π, we define

$$V_\pi(i) = E_\pi\left[\sum_{t=0}^{\infty} \alpha^t C(X_t, a_t) \,|\, X_0 = i\right], \qquad i \geq 0 \qquad (1)$$

where E_π represents the conditional expectation given that the policy π is employed. Since $C(X_t, a_t)$ is just the cost incurred at time t, it follows that $V_\pi(i)$ represents the expected total discounted cost incurred when the policy π is employed and the initial state is i. [Note that (1) is well defined, since costs are bounded and $\alpha < 1$, and also note that the dependence on α is suppressed in the notation $V_\pi(i)$.]

The rationale behind the discount factor is that costs incurred at future dates are less bothersome than costs incurred today, and thus we discount such costs at a rate α per unit of time.

Now, let

$$V_\alpha(i) = \inf_\pi V_\pi(i), \qquad i \geq 0$$

A policy π^* is said to be α-*optimal* if

$$V_{\pi^*}(i) = V_\alpha(i) \qquad \text{for all } i \geq 0$$

Hence, a policy is α-optimal if its expected α-discounted cost is minimal for every initial state.

The following theorem yields a functional equation satisfied by the optimal cost function V_α.

Theorem 6.1

$$V_\alpha(i) = \min_a \left\{ C(i, a) + \alpha \sum_{j=0}^{\infty} P_{ij}(a) V_\alpha(j) \right\}, \qquad i \geq 0 \tag{2}$$

PROOF. Let π be any arbitrary policy, and suppose that π chooses action a at time 0 with probability P_a, $a \in A$. Then,

$$V_\pi(i) = \sum_{a \in A} P_a \left[C(i, a) + \sum_{j=0}^{\infty} P_{ij}(a) W_\pi(j) \right]$$

where $W_\pi(j)$ represents the expected discounted cost incurred from time 1 onward, given that π is being used and the state at time 1 is j. However, it follows that

$$W_\pi(j) \geq \alpha V_\alpha(j)$$

and hence that

$$V_\pi(i) \geq \sum_{a \in A} P_a \left[C(i, a) + \alpha \sum_{j=0}^{\infty} P_{ij}(a) V_\alpha(j) \right]$$

$$\geq \sum_{a \in A} P_a \min_a \left\{ C(i, a) + \alpha \sum_{j=0}^{\infty} P_{ij}(a) V_\alpha(j) \right\} \tag{3}$$

$$= \min_a \left\{ C(i, a) + \alpha \sum_{j=0}^{\infty} P_{ij}(a) V_\alpha(j) \right\}$$

Since π is arbitrary, (3) implies that

$$V_\alpha(i) \geq \min_a \left\{ C(i, a) + \alpha \sum_{j=0}^{\infty} P_{ij}(a) V_\alpha(j) \right\} \tag{4}$$

To go the other way, let a_0 be such that

$$C(i, a_0) + \alpha \sum_{j=0}^{\infty} P_{ij}(a_0)V_\alpha(j) = \min_a \left\{ C(i, a) + \alpha \sum_{j=0}^{\infty} P_{ij}(a)V_\alpha(j) \right\} \quad (5)$$

and let π be the policy that chooses a_0 at time 0; and, if the next state is j, then views the process as originating in state j; and follows a policy π_j, which is such that $V_{\pi_j}(j) \leq V_\alpha(j) + \varepsilon, j \geq 0$. Hence,

$$V_\pi(i) = C(i, a_0) + \alpha \sum_{j=0}^{\infty} P_{ij}(a_0)V_{\pi_j}(j)$$

$$\leq C(i, a_0) + \alpha \sum_{j=0}^{\infty} P_{ij}(a_0)V_\alpha(j) + \alpha\varepsilon$$

which, since $V_\alpha(i) \leq V_\pi(i)$, implies that

$$V_\alpha(i) \leq C(i, a_0) + \alpha \sum_{j=0}^{\infty} P_{ij}(a_0)V_\alpha(j) + \alpha\varepsilon$$

Hence, from (5), we obtain

$$V_\alpha(i) \leq \min_a \left\{ C(i, a) + \alpha \sum_{j=0}^{\infty} P_{ij}(a)V_\alpha(j) \right\} + \alpha\varepsilon \quad (6)$$

and the result follows from (4) and (6), since ε is arbitrary.

Now, as previously noted, a stationary policy is just a function mapping the state space into the action space. That is, the stationary policy f is the policy which, when the process is in state i, selects action $f(i)$.

Let $B(I)$ denote the set of all bounded (real-valued) functions on the state space. Note that (since costs are bounded) $V_\pi \in B(I)$ for all policies π. For any stationary policy f we define the mapping

$$T_f : B(I) \to B(I)$$

in the following manner:

$$(T_f u)(i) = C[i, f(i)] + \alpha \sum_{j=0}^{\infty} P_{ij}[f(i)]u(j) \quad (7)$$

That is, for a function $u \in B(I)$, $T_f u$ is the function whose value at i is given by (7). It is easy to check that $T_f u$ is also bounded and hence is in $B(I)$.

A valuable way of interpreting $T_f u$ is to interpret its value at i as representing the expected cost if we use policy f but we are terminated after one period and incur a final cost $\alpha u(j)$ when the final state is j.

NOTATION. Let $T_f^1 = T_f$, and for $n > 1$ let $T_f^n = T_f(T_f^{n-1})$.

Definitions

For any two functions $u, v \in B(I)$, we say that $u \leq v$ if $u(i) \leq v(i)$ for all $i \geq 0$. Similarly, for $u = v$. Also, for $u_n, u \in B(I)$, we say that $u_n \to u$ if $u_n(i) \to u(i)$ uniformly in i, for $i \geq 0$.

The following lemma yields some important properties of the mappings T_f.

Lemma 6.2

For $u, v \in B(I)$, and f a stationary policy:

(i) $u \leq v \Rightarrow T_f u \leq T_f v$

(ii) $T_f V_f = V_f$

(iii) $T_f^n u \to V_f$ *for all $u \in B(I)$.*

PROOF. Part (i) follows directly from the definition of T_f. Part (ii) is just the statement that

$$V_f(i) = C[i, f(i)] + \alpha \sum_{j=0}^{\infty} P_{ij}[f(i)]V_f(j)$$

which is seen to be true by conditioning on the state at time 1. To prove (iii), we first note that

$(T_f^2 u)(i)$

$$= C[i, f(i)] + \alpha \sum_{j=0}^{\infty} P_{ij}[f(i)](T_f u)(j)$$

$$= C[i, f(i)] + \alpha \sum_{j=0}^{\infty} P_{ij}[f(i)]\left[C[j, f(j)] + \alpha \sum_{k=0}^{\infty} P_{jk}[f(j)]u(k) \right]$$

$$= C[i, f(i)] + \alpha \sum_{j=0}^{\infty} P_{ij}[f(i)]C[j, f(j)] + \alpha^2 \sum_{j=0}^{\infty} \sum_{k=0}^{\infty} P_{ij}[f(i)]P_{jk}[f(j)]u(k)$$

Or, in other words, $T_f^2 u$ represents the expected cost if we use policy f but we are terminated after two periods and incur a final cost $\alpha^2 u$. A simple induction argument then shows that $T_f^n u$ represents the expected cost if we use f for n steps and then incur final cost $\alpha^n u$. Since $\alpha < 1$ and u is bounded, the result follows.

We are now in position to prove the following important theorem.

Theorem 6.3

Let f_α be the stationary policy, which, when the process is in state i, selects the action (or an action) minimizing the right side of (2), i.e., $f_\alpha(i)$ is such that

$$C[i, f_\alpha(i)] + \alpha \sum_{j=0}^{\infty} P_{ij}[f_\alpha(i)]V_\alpha(j) = \min_a \left\{ C(i, a) + \alpha \sum_{j=0}^{\infty} P_{ij}(a)V_\alpha(j) \right\}, \qquad i \geq 0$$

Then

$$V_{f_\alpha}(i) = V_\alpha(i) \qquad \text{for all} \quad i \geq 0$$

and hence f_α is α-optimal.

PROOF. By applying the mapping T_{f_α} to V_α, we obtain

$$(T_{f_\alpha} V_\alpha)(i) = C[i, f_\alpha(i)] + \alpha \sum_{j=0}^{\infty} P_{ij}[f_\alpha(i)]V_\alpha(j)$$

$$= \min_a \left\{ C(i, a) + \alpha \sum_{j=0}^{\infty} P_{ij}(a)V_\alpha(j) \right\}$$

$$= V_\alpha(i)$$

where the last equation follows from Theorem 6.1. Hence,

$$T_{f_\alpha} V_\alpha = V_\alpha$$

which implies that

$$T_{f_\alpha}^2 V_\alpha = T_{f_\alpha}(T_{f_\alpha} V_\alpha) = T_{f_\alpha} V_\alpha = V_\alpha$$

and by induction,

$$T_{f_\alpha}^n V_\alpha = V_\alpha \qquad \text{for all} \quad n$$

Letting $n \to \infty$ and using (iii) of Lemma 6.2 yields the desired result, namely,

$$V_{f_\alpha} = V_\alpha$$

Thus, an α-optimal policy exists, it may be taken to be stationary, and it is determined by the functional equation (2). Hence, if we can determine the optimal expected cost function V_α, then the stationary policy which, when in state i, selects the action (or an action if there is more than one) minimizing $C(i, a) + \alpha \sum_{j=0}^{\infty} P_{ij}(a)V_\alpha(j)$ is α-optimal.

Consider now the following situation. Suppose we have evaulated the expected cost function V_f of some stationary policy f, and we now let f^* denote the stationary policy which when in i selects the action minimizing

$C(i, a) + \alpha \sum_{j=0}^{\infty} P_{ij}(a)V_f(j)$; i.e., $f^*(i)$ is such that

$$C[i, f^*(i)] + \alpha \sum_{j=0}^{\infty} P_{ij}[f^*(i)]V_f(j) = \min_a \left\{ C(i, a) + \alpha \sum P_{ij}(a)V_f(j) \right\}, \qquad i \geq 0 \tag{8}$$

How good is f^* compared with f? The following lemma shows that f^* is at least as good as f. (It will later be shown that f^* is strictly better than f for at least one initial state, or else f is α-optimal.)

Corollary 6.4

$$V_{f^*}(i) \leq V_f(i) \qquad \text{for all } i \geq 0$$

PROOF.

$$(T_{f^*}V_f)(i) = C[i, f^*(i)] + \alpha \sum_{j=0}^{\infty} P_{ij}[f^*(i)]V_f(j)$$

$$\leq C[i, f(i)] + \alpha \sum_{j=0}^{\infty} P_{ij}[f(i)]V_f(j) \tag{9}$$

$$= V_f(i) \tag{10}$$

where (9) follows from the definition of f^*, and (10) follows from Theorem 6.1. Hence,

$$T_{f^*}V_f \leq V_f$$

and applying T_{f^*} to both sides of the above yields by the monotonicity of T_{f^*} [Lemma 6.2 (i)] that

$$T_{f^*}^2 V_f \leq T_{f^*}V_f \leq V_f$$

and by induction we obtain

$$T_{f^*}^n V_f \leq V_f$$

and by letting $n \to \infty$, and using Lemma 6.2 (iii), we obtain the desired result.

The technique of starting with an initial policy and then using Corollary 6.4 to improve it, and then improving this new policy, etc., is known as the *policy improvement algorithm*. We shall have more to say about this in Section 6.

Contraction Mappings

The next question is clearly how do we go about determining V_α? Before dealing with this question, we shall first need some preliminaries.

For any function $u \in B(I)$, let

$$\|u\| = \sup_{i \geq 0} |u(i)|$$

DEFINITION. A mapping $T : B(I) \to B(I)$ is said to be a *contraction mapping* if

$$\|Tu - Tv\| \leq \beta \|u - v\|$$

for some $\beta < 1$, and for all $u, v \in B(I)$. $[u - v$ is the function whose value at i is $u(i) - v(i)$.]

We shall state the following theorem without proof. For a proof see Appendix 1.

Theorem (Contraction Mapping Fixed Point Theorem)

If $T : B(I) \to B(I)$ is a contraction mapping, then there exists a unique function $g \in B(I)$ such that

$$Tg = g$$

Furthermore, for all $u \in B(I)$,

$$T^n u \to g \text{ as } n \to \infty$$

In order to apply this theorem, let us define the mapping $T_\alpha : B(I) \to B(I)$ by

$$(T_\alpha u)(i) = \min_a \left\{ C(i, a) + \alpha \sum_{j=0}^{\infty} P_{ij}(a) u(j) \right\} \tag{11}$$

Note that by Theorem 6.1, it follows that

$$T_\alpha V_\alpha = V_\alpha$$

Thus, if we can show that T_α is a contraction mapping, then it will follow that V_α is the unique solution to (2), and also that V_α may be obtained (at least in the limit) by successively applying T_α to any initial function $u \in B(I)$. This is known as the method of *successive approximations*.

Theorem 6.5

The mapping T_α, as defined by (11), *is a contraction mapping.*

PROOF. For any $u, v \in B(I)$,

$$(T_\alpha u)(i) - (T_\alpha v)(i) = \min_a \left\{ C(i, a) + \alpha \sum_{j=0}^{\infty} P_{ij}(a)u(j) \right\}$$
$$- \min_a \left\{ C(i, a) + \alpha \sum_{j=0}^{\infty} P_{ij}(a)v(j) \right\}$$
$$= \min_a \left\{ C(i, a) + \alpha \sum_{j=0}^{\infty} P_{ij}(a)u(j) \right\}$$
$$- C(i, \bar{a}) - \alpha \sum_{j=0}^{\infty} P_{ij}(\bar{a})v(j) \tag{12}$$

where $\bar{a}$ is such that

$$C(i, \bar{a}) + \alpha \sum_{j=0}^{\infty} P_{ij}(\bar{a})v(j) = \min_a \left\{ C(i, a) + \alpha \sum_{j=0}^{\infty} P_{ij}(a)v(j) \right\}$$

Hence, from (12), we obtain

$$(T_\alpha u)(i) - (T_\alpha v)(i) \le \alpha \sum_{j=0}^{\infty} P_{ij}(\bar{a})u(j) - \alpha \sum_{j=0}^{\infty} P_{ij}(\bar{a})v(j)$$
$$= \alpha \sum_{j=0}^{\infty} P_{ij}(\bar{a})[u(j) - v(j)] \tag{13}$$
$$\le \alpha \sum_{j=0}^{\infty} P_{ij}(\bar{a}) \sup_j [u(j) - v(j)]$$
$$\le \alpha \|u - v\|$$

Thus, from (13), we obtain

$$\sup_{i \ge 0} \{(T_\alpha u)(i) - (T_\alpha v)(i)\} \le \alpha \|u - v\| \tag{14}$$

By reversing the roles of u and v, we obtain

$$\sup_{i \ge 0} \{(T_\alpha v)(i) - (T_\alpha u)(i)\} \le \alpha \|u - v\| \tag{15}$$

and by combining (14) and (15), we obtain

$$\sup_{i \ge 0} |(T_\alpha u)(i) - (T_\alpha v)(i)| \le \alpha \|u - v\|$$

or

$$\|T_\alpha u - T_\alpha v\| \le \alpha \|u - v\|$$

and the result is proven.

As a direct consequence of the above theorem and the contraction mapping fixed point theorem we obtain the following corollary.

Corollary 6.6

V_α is the unique solution to

$$V_\alpha(i) = \min_a \left\{ C(i, a) + \alpha \sum_{j=0}^{\infty} P_{ij}(a) V_\alpha(j) \right\}, \qquad i \geq 0$$

Furthermore, for any $u \in B(I)$,

$$T_\alpha^n u \to V_\alpha \quad \text{as} \quad n \to \infty$$

REMARK 1.   A particularly useful choice is to let u be the zero function, i.e., $u(i) = 0$ for all i. For then, if we let

$$V_\alpha(i, n) = (T_\alpha^n 0)(i)$$

it follows that $V_\alpha(i, n)$ equals the minimal expected discounted cost for an n-period problem. Often important results concerning $V_\alpha(i)$ may be proven by first proving it for $V_\alpha(i, n)$ and then letting $n \to \infty$. For instance, one might show that $V_\alpha(i)$ is monotone in i by first showing that this is true for $V_\alpha(i, n)$. The similarity of the structure of $V_\alpha(i)$ and $V_\alpha(i, n)$ will be further indicated in the next section.

REMARK 2.   Corollary 6.6 also enables us to show that for the policy improvement technique, either the new policy is strictly better than the old one, or else they are both optimal. This follows since if $V_{f^*} = V_f$, then by (8) and Lemma 6.2 (ii) V_f satisfies the optimality Equation (2), and hence by uniqueness $V_f = V_\alpha$.

REMARK 3.   It is also easy to prove that T_f is a contraction mapping. Hence by Lemma 6.2 (ii), V_f is the unique solution to

$$V_f(i) = C[i, f(i)] + \alpha \sum_{j=0}^{\infty} P_{ij}[f(i)] V_f(j)$$

Generalizations

Our results, though proven for a countable state space, easily extend to uncountable state spaces. The assumption of a finite action space is, however, essential.

Also, $C(i, a)$ need not be an actual cost, but it may represent the expected cost if action a is taken when in state i.

6.3. Some Examples

EXAMPLE 1. *A Machine Replacement Model.* Consider a machine which can be in any one of the states 0, 1, 2, ... Suppose that at the beginning of each day, the state of the machine is noted and a decision upon whether or not to replace the machine is made. If the decision to replace is made, then we assume that the machine is instantaneously replaced by a new machine whose state is 0.

The cost of replacing the machine will be denoted by R, and furthermore, we suppose that a maintenance cost $C(i)$ is incurred each day that the machine is in state i.

Also, we let P_{ij} represent the probability that a machine in state i at the beginning of one day will be in state j at the beginning of the next day.

It follows that the above is a two-action Markov decision model in which action 1 is the replacement action and action 2 the nonreplacement action. The one-stage costs and transition probabilities are given by:

$$C(i, 1) = R + C(0), \qquad C(i, 2) = C(i), \qquad i \geq 0$$
$$P_{ij}(1) = P_{0j}, \qquad\qquad P_{ij}(2) = P_{ij}, \qquad i \geq 0$$

Furthermore, we impose the following assumptions on the costs and transition probabilities:

(i) $\{C(i), i \geq 0\}$ is a bounded, increasing sequence
(ii) $\sum_{j=k}^{\infty} P_{ij}$ is an increasing function of i, for each $k \geq 0$.

Hence, (i) asserts that the maintenance cost is an increasing function of the state; and (ii) asserts that the probability of a transition into any block of states $\{k, k + 1, \ldots\}$ is an increasing function of the present state.

In order to determine the structure of the optimal policy, we shall need the following lemma, whose proof is left to the reader.

Lemma 6.7

Assumption (ii) implies that for any increasing function $h(i)$, the function

$$\sum_{j=0}^{\infty} P_{ij} h(j)$$

is also increasing in i.

Lemma 6.8

Under (i) and (ii), $V_\alpha(i)$ is increasing in i.

Proof. Let

$$V_\alpha(i, 1) = \min\{R + C(0); C(i)\}, \qquad i \geq 1$$

and for $n > 1$,

$$V_\alpha(i, n) = \min\left\{R + C(0) + \alpha \sum_{j=0}^{\infty} P_{0j} V_\alpha(j, n-1); \; C(i) + \alpha \sum_{j=0}^{\infty} P_{ij} V_\alpha(j, n-1)\right\}$$

It follows from Assumption (i) that $V_\alpha(i, 1)$ is increasing in i, and if we assume that $V_\alpha(i, n-1)$ is increasing in i, then $V_\alpha(i, n)$ is also increasing by Lemma 6.7. Therefore, it follows by induction that $V_\alpha(i, n)$ is increasing in i for all n, and hence that $V_\alpha(i) = \lim_n V_\alpha(i, n)$ is increasing.

The structure of the optimal policy is now given by the following theorem.

Theorem 6.9

Under Assumptions (i) and (ii), there exists an integer i^*, $i^* \leq \infty$, such that an α-optimal policy replaces for $i > i^*$ and does not replace for $i \leq i^*$.

Proof. By Theorem 6.1, we have

$$V_\alpha(i) = \min\left\{R + C(0) + \alpha \sum_{j=0}^{\infty} P_{0j} V_\alpha(j); C(i) + \alpha \sum_{j=0}^{\infty} P_{ij} V_\alpha(j)\right\}, \qquad i \geq 0 \quad (16)$$

Let

$$i^* = \max\left\{i : C(i) + \alpha \sum_{j=0}^{\infty} P_{ij} V_\alpha(j) \leq R + C(0) + \alpha \sum_{j=0}^{\infty} P_{0j} V_\alpha(j)\right\}$$

Now, by the previous two lemmas, it follows that $C(i) + \alpha \sum_{j=0}^{\infty} P_{ij}V_\alpha(j)$ is increasing in i, and hence by (16),

$$V_\alpha(i) = \begin{cases} C(i) + \alpha \sum_{j=0}^{\infty} P_{ij} V_\alpha(j) & \text{for } i \leq i^* \\[2mm] R + C(0) + \alpha \sum_{j=0}^{\infty} P_{0j} V_\alpha(j) & \text{for } i > i^* \end{cases}$$

The result then follows from Theorem 6.3.

Example 2. *A Quality Control Model.* Consider a machine which can be in one of two states, *good* or *bad*. Suppose that the machine produces an item at the beginning of each day. The item produced is either good (if the machine is good) or bad (if the machine is bad).

We suppose that once the machine is in the bad state, it remains in that state until it is replaced. However, if it is in the good state at the beginning of a day, then with probability γ it will be in the bad state at the beginning of the next day.

We further suppose that after the item is produced we have the option of inspecting the item or not. If the item is inspected and found to be in the bad state, then the machine is instantaneously replaced with a good machine at an additional cost R. Also, the cost of inspecting an item will be denoted by I, and the cost of producing a bad item by C.

We shall say that the process is in state p at time t if p is the posterior probability at t that the machine in use is in the bad state.

It is easily seen that the above is a two-action Markov decision process with an uncountable state space (namely [0, 1], the set of all possible probabilities).

If the state is p and the inspect action is chosen, then we incur an expected cost

$$I + p(C + R)$$

and the next state will be γ, since γ is the probability that a good machine turns bad overnight.

If the state is p and the machine is not inspected, then the expected cost incurred is

$$pC$$

and the next state of the process is $p + (1 - p)\gamma$. Hence, the optimal cost function will satisfy

$$V_\alpha(p) = \min\{I + p(C + R) + \alpha V_\alpha(\gamma); \ pC + \alpha V_\alpha(p + (1 - p)\gamma)\}, \ p \in [0, 1]$$

EXAMPLE 3. *Selling an Asset.* Consider an individual who desires to sell his house. An offer is made at the beginning of each day, and the individual must immediately decide whether or not to accept the offer. Once rejected, the offer is lost. Suppose that the successive offers are independent of each other and take on the value i with probability P_i, for $i = 0, 1, \ldots, N$. Suppose also that a maintenance cost C is incurred each day that the house remains unsold, and that future costs are discounted at a rate α.

If we let the state at time t be the offer at t, then the above is a two-action Markov decision process. (Once an offer is accepted, we suppose that the process goes to state ∞, from which it can never return, and at which all future costs are 0.)

If the state is i, and the offer is accepted, then the one-period cost is $-i$, and if rejected, then the one-period cost is C. Hence,

$$V_\alpha(i) = \min\left\{-i, \ C + \alpha \sum_{j=0}^{N} P_j V_\alpha(j)\right\}, \qquad i = 0, 1, \ldots, N$$

If we let

$$i^* = \min\left\{ i : -i < C + \alpha \sum_{j=0}^{N} P_j V_\alpha(j) \right\}$$

then it follows that the α-optimal policy accepts any offer than greater than or equal to i^*, and rejects all offers less than i^*.

Hence, the structure of the optimal policy is determined. However, by making use of the known structure, it becomes a simple matter to determine the optimal policy (or equivalently, i^*).

Let f_i denote the policy which accepts any offer greater than or equal to i. For this policy, the conditional expected discounted cost given T, the number of rejected offers, is

$$C + \alpha C + \cdots + \alpha^{T-1} C - \alpha^T \frac{\sum_{j=i}^{N} j P_j}{\sum_{j=i}^{N} P_j} = \frac{C(1 - \alpha^T)}{1 - \alpha} - \alpha^T \frac{\sum_{j=i}^{N} j P_j}{\sum_{j=i}^{N} P_j}$$

Since T is geometric with mean $\sum_{j=0}^{i-1} P_j / \sum_{j=i}^{N} P_j$, it follows that the expected discounted cost under f_i is

$$\sum_{j=0}^{N} P_j V_{f_i}(j) = \frac{C \sum_{j=0}^{i-1} P_j - \sum_{j=i}^{N} j P_j}{1 - \alpha \sum_{j=0}^{i-1} P_j}$$

Hence, i^* may be chosen to minimize the right side of the preceding equation.

6.4. Positive Costs, No Discounting

In this section we shall suppose that all costs are nonnegative, that is, that $C(i, a) \geq 0$ for all i, a. We will not assume a discount factor, and we will not require that the costs be bounded.

For any policy π, let

$$V_\pi(i) = E_\pi\left[\sum_{t=0}^{\infty} C(X_t, a_t) \mid X_0 = i \right], \qquad i \geq 0$$

and let

$$V(i) = \inf_\pi V_\pi(i), \qquad i \geq 0$$

It is possible, of course, that $V(i)$ might be infinite; hence the above model is only of interest if the nature of the problem is such that $V(i) < \infty$ for at least some values of i. For otherwise all policies would be optimal.

A policy π^* is said to be *optimal* if

$$V_{\pi^*}(i) = V(i), \qquad \text{for all } i \geq 0$$

In exactly the same manner as in Section 6.2, we may prove

Theorem 6.10

$$V(i) = \min_a \left\{ C(i, a) + \sum_{j=0}^{\infty} P_{ij}(a)V(j) \right\}, \qquad i \geq 0 \qquad (17)$$

Let us now denote by $N(I)$, the set of all nonnegative (possibly infinite-valued) functions on the state space; and for any stationary policy f we define the mapping

$$T_f : N(I) \to N(I)$$

by

$$(T_f u)(i) = C[i, f(i)] + \sum_{j=0}^{\infty} P_{ij}[f(i)]u(j)$$

Hence, $T_f u$ is the expected cost if we use policy f, but we are terminated after one period and incur a final cost $u(j)$ when the final state is j.

Analogously to Lemma 6.2, we have

Lemma 6.11

For $u, v \in N(I)$, and f a stationary policy:

(i) $u \leq v \Rightarrow T_f u \leq T_f v$
(ii) $T_f V_f = V_f$
(iii) $(T_f^n 0)(i) \to V_f(i)$ *as $n \to \infty$ for each i, where 0 represents the function which is identically zero.*

The proof is similar to the one given for Lemma 6.2 and is left as an exercise. Note, however, that (iii) is only true for the zero function and not for any $u \in B(I)$; also, the convergence in (iii) is pointwise rather than uniform. This is true because we are not assuming a discount factor. For a discount function α, the final cost is $\alpha^n u$, which uniformly goes to zero if $u \in B(I)$. Without discounting, the only way that one can be assured that the final cost goes to zero is to let it be zero. [Hence $(T_f^n 0)i \to V_f(i)$.]

Theorem 6.12

Let f_1 be the stationary policy which, when the process is in state i, selects the action (or an action) minimizing the right side of (17). Then

$$V_{f_1}(i) = V(i), \qquad for\ all\ i \geq 0$$

and hence f_1 is optimal.

PROOF. By applying T_{f_1} to V, we obtain

$$(T_{f_1}V)(i) = C[i, f_1(i)] + \sum_{j=0}^{\infty} P_{ij}[f_1(i)]V(j)$$

$$= \min_a \left\{ C(i, a) + \sum_{j=0}^{\infty} P_{ij}(a)V(j) \right\}$$

$$= V(i)$$

Hence,

$$T_{f_1}V = V$$

Now, $C(i, a) \geq 0$, which implies that $V \geq 0$, and hence by the monotonicity of T_{f_1}, we obtain

$$T_{f_1}0 \leq T_{f_1}V = V$$

and by successively applying T_{f_1} and making use of its monotonicity property, we obtain

$$T_{f_1}^n 0 \leq V$$

Now, by letting $n \to \infty$ and using Lemma 6.11 (iii), we arrive at

$$V_{f_1} \leq V$$

which, since $V \leq V_{f_1}$ by definition, yields the desired result.

Hence, an optimal policy exists and is determined by the functional equation (17). However, we no longer have any contraction mappings and as a result V need not be the unique solution of the functional equation (17). Also, the method of successive approximations is no longer available. The generalization to an arbitrary state space is, as in the discount case, easily accomplished.

6.5. Applications: Optimal Stopping and Sequential Analysis

Optimal Stopping Problems

Consider a process with states $0, 1, 2, \ldots$, in which when in state i we may either stop (action 1) and receive a terminal reward $R(i)$, or (action 2) pay a cost $C(i)$ and go to the next state according to the transition probabilities P_{ij}, $i, j \geq 0$.

If we say that the process goes to state ∞ when the stop action is chosen, then the above is a two action Markov decision process having

$$
\begin{array}{ll}
C(i, 1) = -R(i) & i = 0, 1, 2, \ldots \\
C(i, 2) = C(i) & i = 0, 1, 2, \ldots \\
C(\infty, \cdot) = 0 & \\
P_{i\infty}(1) = 1 & i = 0, 1, 2, \ldots \\
P_{ij}(2) = P_{ij} & i, j = 0, 1, 2, \ldots \\
P_{\infty\infty}(\cdot) = 1 &
\end{array}
$$

We shall suppose that

(i) $\inf\limits_{i \geq 0} C(i) > 0$

and

(ii) $\sup\limits_{i} R(i) < \infty$.

We cannot immediately apply the results of the previous section to this process, since it is not the case that all costs are nonnegative. [Note that the reward $R(i)$ is interpreted as a negative cost.] However, we may transform this process into an equivalent one, for which the results of the previous section do apply, in the following manner. We let $R = \sup_i R(i)$, and consider a related process which is such that when in state i, we may either stop and pay a terminal cost $R - R(i)$, or we may pay a cost $C(i)$ and go to the next state according to P_{ij}, $i, j \geq 0$.

For any policy π, let $V_\pi(\ )$ denote the expected cost with respect to the original process when π is used, and let $V'_\pi(\ )$ denote the expected cost with respect to the related process when π is used. It is easy to see that for any policy π, which stops in finite expected time, we have

$$
V'_\pi(i) = V_\pi(i) + R \qquad i = 0, 1, 2, \ldots
$$

However, these are the only policies we need consider, since by Assumption (i), any policy π which does not stop in finite expected time has $V_\pi(i) = V'_\pi(i) = \infty$. Hence, any policy which is optimal for the original process is optimal for the related process, and vice versa.

However, the related process is a Markov decision process with nonnegative costs, and thus, by the results of the previous section, an optimal policy exists and the optimal cost function $V'(i)$ satisfies

$$
V'(i) = \min\left\{ R - R(i); \ C(i) + \sum_{j=0}^{\infty} P_{ij} V'(j) \right\}, \qquad i = 0, 1, 2, \ldots
$$

Also, the policy which chooses the minimizing actions is optimal.

Or, stating these results in terms of the optimal cost function $V(i)$ $(= V'(i) - R)$ of the original process, we obtain

$$V(i) = \min\left\{-R(i); C(i) + \sum_{j=0}^{\infty} P_{ij} V(j)\right\}, \qquad i = 0, 1, 2, \ldots$$

and the policy which when in state i chooses the action minimizing the right side of the above is optimal.

Let

$$V_0(i) = -R(i)$$

and for $n > 0$,

$$V_n(i) = \min\left\{-R(i); C(i) + \sum_{j=0}^{\infty} P_{ij} V_{n-1}(j)\right\}, \qquad i = 0, 1, 2, \ldots$$

In other words, $V_n(i)$ is the minimal expected cost if we start in i, and if we are allowed to go at most n stages before stopping. From this interpretation it follows that, for all i, n,

$$V_n(i) \geq V_{n+1}(i) \geq V(i)$$

Hence, $\lim_{n \to \infty} V_n(i) \geq V(i)$. We shall say that the process is *stable* if

$$\lim_{n \to \infty} V_n(i) = V(i), \qquad i = 0, 1, 2, \ldots$$

The following theorem not only shows that conditions (i) and (ii) ensure stability, but it also gives bounds on how quickly $V_n(i)$ converges to $V(i)$. Let $R = \sup_i R(i)$, and $C = \inf_i C(i)$.

Theorem 6.13

Assuming conditions (i) and (ii),

$$V_n(i) - V(i) \leq \frac{(R - C)[R - R(i)]}{(n + 1)C} \qquad \text{for all } n, \text{ all } i$$

PROOF. Let f be an optimal policy, and let T denote the random time at which f stops. Also, let f_n be the policy which chooses the same actions as f at times $0, 1, \ldots n - 1$, but which stops at time n (if it had not previously done so). Then,

$$V(i) = V_f(i) = E_f[X \mid T \leq n]P\{T \leq n\} + E_f[X \mid T > n]P\{T > n\} \quad (18)$$

and

$$V_n(i) \leq V_{f_n}(i) = E_f[X \mid T \leq n]P\{T \leq n\} + E_{f_n}[X \mid T > n]P\{T > n\}$$

where X denotes the total cost incurred and everything is understood to be conditional on $X_0 = i$. Thus,

$$V_n(i) - V(i) \le [E_{f_n}(X \mid T > n) - E_f(X \mid T > n)]P\{T > n\}$$
$$\le (R - C)P\{T > n\}$$

To obtain a bound on $P\{T > n\}$, we note by (18) that

$$-R(i) \ge V(i) \ge -R\,P\{T \le n\} + (-R + (n + 1)C)P\{T > n\}$$
$$= -R + (n + 1)CP\{T > n\}$$

or

$$P\{T > n\} \le \frac{R - R(i)}{(n + 1)C}$$

and the desired result follows.

Now, let

$$B = \left\{i : -R(i) \le C(i) - \sum_{j=0}^{\infty} P_{ij}R(j)\right\}$$
$$= \left\{i : R(i) \ge \sum_{j=0}^{\infty} P_{ij}R(j) - C(i)\right\}$$

Hence, B represents the set of states for which stopping is at least as good as continuing for exactly one more period and then stopping.

Theorem 6.14

If the process is stable, and if $P_{ij} = 0$ for $i \in B$, $j \notin B$, then the optimal policy stops at i if and only if $i \in B$.

PROOF.   We shall show that $V_n(i) = -R(i)$ for all $i \in B$, all n. It follows trivially for $n = 0$, so suppose it for $n - 1$. Then, for $i \in B$,

$$V_n(i) = \min\left\{-R(i); C(i) + \sum_{j=0}^{\infty} P_{ij}V_{n-1}(j)\right\}$$
$$= \min\left\{-R(i); C(i) + \sum_{j \in B} P_{ij}V_{n-1}(j)\right\}$$
$$= \min\left\{-R(i); C(i) - \sum_{j \in B} P_{ij}R(j)\right\}$$
$$= -R(i)$$

Hence, $V_n(i) = -R(i)$ for all $i \in B$, all n. By letting $n \to \infty$ and using the stability hypothesis, we obtain

$$V(i) = -R(i) \qquad \text{for } i \in B$$

Now, for $i \notin B$, the policy which continues for exactly one stage and then stops has an expected cost

$$C(i) - \sum_{j=0}^{\infty} P_{ij} R(j)$$

which is strictly less than $-R(i)$ (since $i \notin B$). Hence,

$$V(i) \begin{cases} = -R(i) & \text{for } i \in B \\ < -R(i) & \text{for } i \notin B \end{cases}$$

and the result follows.

REMARK. Since the policy defined in Theorem 6.14 compares stopping immediately with stopping after one period, it is called a *one-stage lookahead policy*. Also, if the conditions of Theorem 6.14 holds, then we are said to be in the monotone case. Hence, the one-stage lookahead policy is optimal in the monotone case.

EXAMPLE 4. *A House Selling Example.* An example illustrating the use of Theorem 6.14 is the following. Suppose an individual wants to sell his house and an offer comes in at the beginning of each day. As in the example of Section 6.3, we suppose that the successive offers are independent and an offer is j with probability P_j, $j = 0, 1, \ldots, N$. We suppose, however, that any offer not immediately accepted is not lost but may be accepted at any later date. Also, a maintenance cost C is incurred for each day that the house remains unsold.

The state at time t will be the largest offer received up to t (including the offer at t). Hence,

$$P_{ij} = \begin{cases} 0 & j < i \\ \displaystyle\sum_{k=0}^{i} P_k & j = i \\ P_j & j > i \end{cases}$$

and thus,

$$\begin{aligned} B &= \left\{ i : -i \le C - i \sum_{k=0}^{i} P_k - \sum_{j=i+1}^{N} j P_j \right\} \\ &= \left\{ i : C \ge \sum_{j=i+1}^{N} j P_j - i \sum_{k=i+1}^{N} P_k \right\} \\ &= \left\{ i : C \ge \sum_{j=i+1}^{N} (j - i) P_j \right\} \end{aligned} \tag{19}$$

Since the right side of (19) is clearly decreasing in i, it follows that

$$B = \{i^*, i^* + 1, \ldots, N\}$$

where

$$i^* = \min\left\{i : C \geq \sum_{j=i+1}^{N} (j - i)P_j\right\}$$

Hence, $P_{ij} = 0$ for $i \in B$, $j \in B$, and thus the policy which accepts the first offer that is at least i^* is optimal.

REMARK. Consider the same problem with the exception that once an offer is rejected, it is no longer available. Clearly, the optimal cost for this problem is no less than the optimal cost for the process we have just considered. Thus, since the optimal policy given above is a legitimate policy for this new problem (as it never accepts an old offer), it follows that this policy is also optimal for this new problem. (Compare this result with the one derived in Section 6.3.)

Sequential Analysis

Let $Y_1, Y_2, \ldots$ be a sequence of independent and identically distributed random variables. Suppose that we know that the probability density function of the Y_i's is either f_0 or f_1 and that we are trying to decide upon one of these.

At time t, after observing $Y_1, Y_2, \ldots, Y_t$, we may either stop observing and choose ei r f_0 or f_1, or we may pay a cost C and observe Y_{t+1}. If we stop observing and make a choice, then we incur a cost 0 if our choice is correct and a cost $L(> 0)$ if it is incorrect.

We shall suppose that we are given an initial probability p_0 that the true density is f_0; and we shall say that the state at t is p if p is the posterior probability at t that f_0 is the true density.

The above is easily seen to be a three action Markov decision process with nonnegative costs and an uncountable state space (namely, [0, 1]).

If the state is p and we stop and choose f_0, then our expected cost is

$$(1 - p)L$$

and if we stop and choose f_1, then it is

$$pL$$

If we take another observation when in state p, then the value observed will be x with probability (density)

$$pf_0(x) + (1 - p)f_1(x)$$

and if the value observed is x, then the next state is

$$X_{t+1} = \frac{pf_0(x)}{pf_0(x) + (1-p)f_1(x)}$$

Hence, the optimal cost function $V(p)$ satisfies

$$V(p) = \min\left\{(1-p)L, pL, C + \int_{-\infty}^{\infty} V\left(\frac{pf_0(x)}{pf_0(x) + (1-p)f_1(x)}\right)\right.$$
$$\left. \times [pf_0(x) + (1-p)f_1(x)] \, dx\right\}, \qquad p \in [0, 1] \tag{20}$$

It is convenient to define a subclass of the class of all policies, and we let Δ denote the class of policies whose action at t is a function only of $Y_1, Y_2, \ldots, Y_t$ and not of the initial state p_0. Since the posterior probability of f_0 at t is a function only of p_0 and $Y_1, \ldots, Y_t$, it follows that given p_0, the optimal policy may be described solely in terms of $Y_1, Y_2, \ldots, Y_t$. Hence,

$$V(p) = \min_{\pi \in \Delta} V_\pi(p), \qquad \text{for each } p \in [0, 1] \tag{21}$$

(Note that the above does not assert that Δ contains an optimal policy. It states that for each p there exists $\pi_p \in \Delta$ such that $V(p) = V_{\pi_p}(p)$, and not that there is a policy $\pi \in \Delta$ that is optimal for all initial p.)

By using (21), we may prove

Lemma 6.15

$V(p)$ *is a concave function of* p.

PROOF. For $\lambda \in (0, 1)$, we have by (21) that

$$V[\lambda p_1 + (1-\lambda)p_2] = \min_{\pi \in \Delta} V_\pi(\lambda p_1 + (1-\lambda)p_2)$$

However, since policies in Δ are independent of the initial probability, it follows from the interpretation of the state as a probability that for $\pi \in \Delta$,

$$V_\pi(\lambda p_1 + (1-\lambda)p_2) = \lambda V_\pi(p_1) + (1-\lambda)V_\pi(p_2)$$

Hence,

$$V(\lambda p_1 + (1-\lambda)p_2) = \min_{\pi \in \Delta}\{\lambda V_\pi(p_1) + (1-\lambda)V_\pi(p_2)\}$$
$$\geq \min_{\pi \in \Delta} \lambda V_\pi(p_1) + \min_{\pi \in \Delta}(1-\lambda)V_\pi(p_2)$$
$$= \lambda V(p_1) + (1-\lambda)V(p_2)$$

and the lemma is proven.

The structure of the optimal policy is given in the next theorem.

Theorem 6.16

There exists numbers p^*, p^{**}, where $p^* < p^{**}$, such that when the state is p, the optimal policy stops and chooses f_0 if $p > p^{**}$, stops and chooses f_1 if $p < p^*$, and continues otherwise (see Figure 6.1).

Figure 6.1. Structure of Optimal Policy

PROOF. Suppose p_1 and p_2 are such that

$$V(p_i) = (1 - p_i)L, \qquad i = 1, 2$$

Then, for any $p = \lambda p_1 + (1 - \lambda)p_2$ where $\lambda \in [0, 1]$, we have, by Lemma 6.15, that

$$V(p) \geq \lambda V(p_1) + (1 - \lambda)V(p_2) = (1 - p)L$$

However, by (20),

$$V(p) \leq (1 - p)L$$

and thus,

$$V(p) = (1 - p)L$$

Hence, $\{p : V(p) = (1 - p)L\}$ is an interval. Also, it contains the point $p = 1$, since $(1 - p)L = 0$ at $p = 1$. From this, it follows that it is optimal to stop and choose f_0 whenever the state p is larger than some p^{**}. Similar comments hold for $\{p : V(p) = pL\}$ and the theorem follows.

6.6 Expected Average Cost Criterion—Introduction and Counterexamples

In this section we shall once again suppose that costs are bounded, and we shall attempt to minimize the long-run expected average cost per unit time. For any policy π, we define

$$\phi_\pi(i) = \lim_{n \to \infty} E_\pi \frac{\left[\sum_{t=0}^{n} C(X_t, a_t) \mid X_0 = i\right]}{n + 1} \tag{22}$$

where, if the limit in (22) does not exist, then we agree to take the lim sup.

Hence, $\phi_\pi(i)$ represents the average expected cost when π is employed and the initial state is i. We shall say that the policy π^* is *average cost optimal* if

$$\phi_{\pi^*}(i) = \min_\pi \phi_\pi(i) \qquad \text{for all } i$$

The first question we shall consider is whether an optimal policy need exist. We shall answer this question by our first counterexample.

COUNTEREXAMPLE 1. The state space will be denoted by the set $\{1, 1', 2, 2', 3, 3', \ldots\}$. There are two actions, and the transition probabilities are given by

$$P_{ii+1}(1) = P_{ii'}(2) = 1$$
$$P_{i'i'}(1) = P_{i'i'}(2) = 1 \qquad i = 1, 2, \ldots$$

The costs depend only on the state and are given by

$$C(i, \cdot) = 1$$
$$C(i', \cdot) = 1/i \qquad i = 1, 2, 3, \ldots$$

In other words, when in state i we may either pay 1 unit and go to $i + 1$, or we may go to i' and as a result pay $1/i$ every day from that time on.

Suppose that $X_0 = 1$, and let π be any policy. Then, there are two cases.

Case 1: With probability 1, π always chooses action 1. In this case,

$$\phi_\pi(1) = 1 > 0$$

Case 2: With positive probability, π chooses action 2 at some time.

In this case, it follows that for some n there exists positive probability $P_n > 0$ that π takes action 2 when in state n. Letting $\bar{n}$ be the smallest n having this property, it easily follows that

$$\phi_\pi(1) \geq \frac{P_{\bar{n}}}{\bar{n}} > 0.$$

Hence, in either case, $\phi_\pi(1) > 0$. However, by choosing action 1 long enough and then choosing action 2, we may make our average cost as close to zero as we desire. Thus, an optimal policy does not exist.

The second question we shall consider is whether we may restrict attention to the class of stationary policies. For example, in the above, while an optimal policy does not exist, it does follow that for any policy there is a stationary policy which does at least as well.

Again, we shall answer this question in the negative by presenting a counterexample.

COUNTEREXAMPLE 2. Let the state space be the positive integers 1, 2, 3, ..., and suppose that there are two actions. The costs and transition probabilities are given as follows:

$$P_{ii+1}(1) = 1 = P_{ii}(2)$$

$$C(i, 1) = 1$$
$$C(i, 2) = 1/i$$
$$i = 1, 2, \ldots$$

In other words, when in state i we may either pay 1 unit and go to $i + 1$, or we may pay $1/i$ units and remain in i.

Suppose $X_0 = 1$, and let π be any stationary policy. There are two cases.

Case 1: π always chooses action 1.

In this case, $\phi_\pi(1) = 1$.

Case 2: π chooses action 2 for the first time at state n.

In this case, the process will go from state 1 to 2 to 3 to n. However, when in state n, action 2 is chosen and thus the process will never leave state n; as a result, $\phi_\pi(1) = 1/n > 0$.

Hence, for any stationary policy π,

$$\phi_\pi(1) > 0$$

Now let π^* be the nonstationary policy which, when the process first enters state i, chooses action 2, i consecutive times, and then chooses action 1. Since the initial state is $X_0 = 1$, it follows that the successive costs incurred under π^* will be

$$1, 1, 1/2, 1/2, 1, 1/3, 1/3, 1/3, 1, 1/4, 1/4, 1/4, 1/4, 1, 1/5, \ldots$$

The average cost of such a sequence is easily seen to be 0, and thus

$$\phi_{\pi^*}(1) = 0$$

Hence, the nonstationary policy π^* is better than every stationary policy.

However, in our definition of a stationary policy, we have not allowed for the possibility of randomization. Let us, therefore, define a randomized stationary policy to be a policy for which the actions chosen, while only depending on the present state, may be random. For example, when in state 1, we may choose action 1 with probability 1/2 and action 2 with probability 1/2.

Can we restrict attention to the randomized stationary policies? Let us first reconsider Counterexample 2. Suppose we let π be the randomized stationary policy which, when in state i, selects action 2 with probability $i/i+1$ and action 1 with probability $1/i+1$. Hence, if we employ π, then when the process first enters state i, the expected number of times that action 2 will be consecutively chosen is i. Hence, it seems reasonable that π will act somewhat like the nonstationary optimal policy π^*, and indeed it may be shown that the average cost under π is zero.

Unfortunately, though the above gives us some hope, it turns out that we cannot, in general, restrict attention even to the randomized stationary policies. An example has been given by Fisher and Ross (see [6]), in which an optimal nonstationary policy is shown to be better than every randomized stationary policy. (This example may be found in Appendix 2.)

We shall not pursue the interesting problem of attempting to find the smallest class of *good* policies,† but instead we shall attempt to determine conditions under which opimal stationary policies exist.

6.7. Expected Average Cost Criterion—Results

We begin this section with a theorem.

Theorem 6.17

If there exists a bounded function $h(i)$, $i = 0, 1, 2, \ldots$, *and a constant* g *such that*

$$g + h(i) = \min_a \left\{ C(i, a) + \sum_{j=0}^{\infty} P_{ij}(a) h(j) \right\}, \qquad i \geq 0 \qquad (23)$$

then there exists a stationary policy π^* *such that*

$$g = \phi_{\pi^*}(i) = \min_{\pi} \phi_{\pi}(i), \qquad \text{for all } i \geq 0$$

and π^* *is any policy which, for each* i, *prescribes an action which minimizes the right-side of* (23).

PROOF. Let $H_t = (X_0, a_0, \ldots, X_t, a_t)$ denote the history of the process up to time t. For any policy π,

$$E_\pi \left\{ \sum_{t=1}^{n} [h(X_t) - E_\pi(h(X_t) \mid H_{t-1})] \right\} = 0$$

But,

$$E_\pi[h(X_t) \mid H_{t-1}] = \sum_{j=0}^{\infty} h(j) P_{X_{t-1}j}(a_{t-1})$$

$$= C(X_{t-1}, a_{t-1}) + \sum_{j=0}^{\infty} h(j) P_{X_{t-1}j}(a_{t-1}) - C(X_{t-1}, a_{t-1})$$

$$\geq \min_a \left\{ C(X_{t-1}, a) + \sum_{j=0}^{\infty} h(j) P_{X_{t-1}j}(a) \right\} - C(X_{t-1}, a_{t-1})$$

$$= g + h(X_{t-1}) - C(X_{t-1}, a_{t-1})$$

† For some partial results in this direction, see the paper by Derman and Strauch [5].

with equality for π^* since π^* is defined to take the minimizing action. Hence

$$0 \leq E_\pi \left\{ \sum_{t=1}^{n} [h(X_t) - g - h(X_{t-1}) + C(X_{t-1}, a_{t-1})] \right\}$$

or

$$g \leq E_\pi \frac{h(X_n)}{n} - E_\pi \frac{h(X_0)}{n} + E_\pi \frac{\sum_{t=1}^{n} C(X_{t-1}, a_{t-1})}{n}$$

with equality for π^*. Letting $n \to \infty$ and using the fact that h is bounded, we have that

$$g \leq \phi_\pi(X_0)$$

with equality for π^*, and for all possible values of X_0. Hence, the desired result is proven.

Therefore, if the conditions of Theorem 6.17 are satisfied, then a stationary optimal policy exists and may be characterized by the functional equation (23). This, however, immediately raises two questions. The first is why such a theorem should indeed be true. For, as it now stands, Theorem 6.17 seems in no way obvious or intuitive. And second, when are the conditions of Theorem 6.17 satisfied? In what follows, we shall attempt both to consider Theorem 6.17 intuitively and also to determine sufficient conditions for it to be of use.

For the discounted cost criterion, future periods are discounted at a rate α, while in the average cost case, all periods are given equal weight. Hence, it seems reasonable that under certain conditions, the average cost case should be in some sense a limit of the discount case as the discount factor approaches unity.

Now, the α-optimal cost function $V_\alpha(i)$ satisfies

$$V_\alpha(i) = \min_a \left\{ C(i, a) + \alpha \sum_{j=0}^{\infty} P_{ij}(a) V_\alpha(j) \right\}, \qquad i \geq 0 \qquad (24)$$

and the α-optimal policy selects the minimizing actions. One possible means of obtaining an optimal average cost policy might thus be to choose the actions minimizing

$$\lim_{\alpha \to 1} \left[C(i, a) + \alpha \sum_{j=0}^{\infty} P_{ij}(a) V_\alpha(j) \right]$$

However, this limit need not exist and indeed would often be infinite for all actions. As a result, this direct approach is not fruitful.

Let us now consider the following indirect approach to the problem. Fix some state—say state 0—and let us define

$$h_\alpha(i) = V_\alpha(i) - V_\alpha(0)$$

to be the α-cost of state i relative to state 0. Then, from (24), we obtain

$$(1 - \alpha)V_\alpha(0) + h_\alpha(i) = \min_a\left\{C(i, a) + \alpha \sum_{j=0}^{\infty} P_{ij}(a)h_\alpha(j)\right\} \tag{25}$$

Also, it follows that the policy that chooses actions so as to minimize the right side of (25) is an α-optimal policy. Now, if for some sequence $\alpha_n \to 1$, $h_{\alpha_n}(j)$ converges to a function $h(j)$ and $(1 - \alpha_n)V_{\alpha_n}(0)$ converges to a constant g, then we obtain from (25) that

$$g + h(i) = \min_a\left\{C(i, a) + \sum_{j=0}^{\infty} P_{ij}(a)h(j)\right\}$$

where we have assumed that the interchange of summation and limit is justified. Also, it seems reasonable that the policy which chooses the minimizing actions will be average cost optimal, i.e., that Theorem 6.17 will be true.

Now, formally we prove

Theorem 6.18

If there exists an $N < \infty$ such that

$$|V_\alpha(i) - V_\alpha(0)| < N \quad \text{for all } \alpha, \text{ all } i$$

then:

    (i) *There exists a bounded function $h(i)$ and a constant g satisfying (23);*
    (ii) *For some sequence $\alpha_n \to 1$, $h(i) = \lim_{n \to \infty} V_{\alpha_n}((i) - V_{\alpha_n}(0))$;*
    (iii) *$\lim_{\alpha \to 1}(1 - \alpha)V_\alpha(0) = g$.*

PROOF. By assumption, $h_\alpha(i) = V_\alpha(i) - V_\alpha(0)$ is uniformly bounded in α and i. Hence, since the state space is countable, we can, by Cauchy's diagonalization method, get a sequence $\alpha_n \to 1$ such that $\lim_{n \to \infty} h_{\alpha_n}(i) \equiv h(i)$ exists for all i. Also, since costs are bounded, it follows that $(1 - \alpha_n)V_{\alpha_n}(0)$ is bounded and thus there exists a subsequence $\{\alpha'_n\}$ of $\{\alpha_n\}$ such that $\lim_{n \to \infty}(1 - \alpha'_u)V_{\alpha'_n}(0) \equiv g$ exists. Now, from (25) we have

$$(1 - \alpha'_n)V_{\alpha'_n}(0) + h_{\alpha'_n}(i) = \min_a\left\{C(i, a) + \alpha'_n \sum_{j=0}^{\infty} P_{ij}(a)h_{\alpha'_n}(j)\right\}$$

Hence, the result (i) follows by letting $n \to \infty$, and noting that the boundedness of $h_{\alpha'_n}(i)$ implies, by Lebesque's bounded convergence theorem, that

$$\sum_{j=0}^{\infty} P_{ij}(a)h_{\alpha'_n}(j) \to \sum_{j=0}^{\infty} P_{ij}(a)h(j)$$

To prove (iii), we note that since $(1 - \alpha)V_\alpha(0)$ is bounded, it follows that for any sequence $\alpha_n \to 1$ there is a subsequence α'_n such that

$$\lim_{n \to \infty} (1 - \alpha'_n)V_{\alpha'_n}(0)$$

exists. By the proof of (i) it follows that this limit must be g. Hence,

$$g = \lim_{\alpha \to 1} (1 - \alpha)V_\alpha(0)$$

and the proof is complete.

REMARK.   From (ii) it follows that $h(i)$ inherits the structural form of $V_\alpha(i)$. For example, if $V_\alpha(i)$ is increasing (is convex) then it follows that $h(i)$ is increasing (is convex).

AN EXAMPLE.   Consider the machine replacement example of Section 6.3. By (16), we have that

$$V_\alpha(0) = C(0) + \alpha \sum_{j=0}^{\infty} P_{0j} V_\alpha(j)$$

and also that

$$V_\alpha(i) \leq R + C(0) + \alpha \sum_{j=0}^{\infty} P_{0j} V_\alpha(j) = R + V_\alpha(0)$$

As a result, since $V_\alpha(i)$ is increasing in i (Lemma 6.8), it follows that

$$|V_\alpha(i) - V_\alpha(0)| \leq R$$

Hence, there exists a constant g and an increasing function $h(i)$ such that

$$g + h(i) = \min\left\{R + C(0) + \sum_{j=0}^{\infty} P_{0j} h(j); C(i) + \sum_{j=0}^{\infty} P_{ij} h(j)\right\}, \qquad i \geq 0$$

and the policy which chooses the minimizing actions is average cost optimal. If we let

$$i^* = \max\left\{i : C(i) + \sum_{j=0}^{\infty} P_{ij} h(j) \leq R + C(0) + \sum_{j=0}^{\infty} P_{0j} h(j)\right\}$$

then it follows, as in the proof of Theorem 6.9, that the average cost optimal policy has the same structure as the α-optimal policy.

The following theorem gives a sufficient condition for $V_\alpha(i) - V_\alpha(0)$ to be uniformly bounded.

Theorem 6.19

If for some state—call it 0—there is a constant $N < \infty$ such that

$$M_{i0}(f_\alpha) < N, \qquad \text{for all } i, \text{ all } \alpha$$

then $V_\alpha(i) - V_\alpha(0)$ is uniformly bounded, where $M_{i0}(f_\alpha)$ is the mean recurrence time to go from state i to state 0 when using the α-optimal policy f_α.

Before proving Theorem 6.19, we state without proof the following theorem, which is known as *Jensen's inequality*.

Theorem (Jensen's Inequality)

  If $g(x)$ is a convex function and X a random variable, then

$$Eg(X) \geq g(EX)$$

provided the expectations exist.

PROOF OF THEOREM 6.19.   We first note that we may, without loss of generality, assume that all costs are nonnegative. This is true since costs are bounded, and adding a constant to all the costs $C(i, a)$ will affect all rules identically. Let

$$T = \min\{t : X_t = 0\}$$

Then,

$$V_\alpha(i) = E_{f_\alpha} \sum_{n=0}^{T-1} C(X_n, a_n)\alpha^n + E_{f_\alpha} \sum_{n=T}^{\infty} C(X_n, a_n)\alpha^n \tag{26}$$

where all expectations are understood to be conditional on $X_0 = i$. Therefore,

$$V_\alpha(i) \leq ME_{f_\alpha} T + V_\alpha(0)E_{f_\alpha}[\alpha^T]$$
$$\leq MN + V_\alpha(0)$$

where M is the bound on costs. To get the inequality in the other direction, we note by (26) that

$$V_\alpha(i) \geq V_\alpha(0)E_{f_\alpha}[\alpha^T]$$

or equivalently,

$$V_\alpha(0) \leq V_\alpha(i) + (1 - E_{f_\alpha}[\alpha^T])V_\alpha(0)$$

Now, $V_\alpha(0) \leq M/1 - \alpha$ and $E\alpha^T \geq \alpha^{ET} \geq \alpha^N$ by Jensen's inequality. Hence,

$$V_\alpha(0) \leq V_\alpha(i) + (1 - \alpha^N) \frac{M}{1 - \alpha} < V_\alpha(i) + MN$$

and the result follows.

As a corollary to Theorem 6.19, we have

Corollary 6.20

If the state space is finite and every stationary policy gives rise to an irreducible Markov chain, then $V_\alpha(i) - V_\alpha(0)$ is uniformly bounded, and hence the conditions of Theorem 6.17 are satisfied.

PROOF. Since a finite Markov chain cannot be null recurrent, it follows that $M_{i0}(f) < \infty$ for all i, and all stationary policies f. The result then follows from Theorem 6.19, since there are only a finite number of stationary policies as well as a finite number of states.

REMARK. Actually, it is not necessary that every stationary policy gives rise to an irreducible Markov chain. It is clearly sufficient if there is some state, say 0, which is accessible from every other state regardless of which α-optimal policy is being used.

We shall end this section by showing how, in a special case, the average cost criterion may be reduced to a discounted cost criterion. We shall need the following assumption.

ASSUMPTION. There is a state, call it 0, and $\beta > 0$, such that

$$P_{i0}(a) \geq \beta \qquad \text{for all} \quad i, \quad \text{all } a$$

For any process satisfying the above assumption, consider a new process with identical state and action spaces, with identical costs, but with transition probabilities now given by

$$P'_{ij}(a) = \begin{cases} \dfrac{P_{ij}(a)}{1 - \beta} & j \neq 0 \\[2ex] \dfrac{P_{i0}(a) - \beta}{1 - \beta} & j = 0 \end{cases} \tag{27}$$

Also, let $V'_{1-\beta}(i)$ be the $(1 - \beta)$-optimal cost function for this new process. Now, letting

$$h'(i) = V'_{1-\beta}(i) - V'_{1-\beta}(0)$$

we have by (25) that

$$\begin{aligned} \beta V'_\beta(0) + h'(i) &= \min_a \left\{ C(i, a) + (1 - \beta) \sum_{j=0}^\infty P'_{ij}(a) h'(j) \right\} \\ &= \min_a \left\{ C(i, a) + \sum_{j=0}^\infty P_{ij}(a) h'(j) \right\} \end{aligned} \tag{28}$$

where the last equation follows from (27) since $h'(0) = 0$. Hence, the conditions of Theorem 6.17 are satisfied. It follows that $g = \beta V'_\beta(0)$ and the average cost optimal policy is the one which selects the actions which minimize the right side of (28). But it is easily seen that this is precisely the $(1 - \beta)$-optimal policy for the new process. Hence, summing up, the average cost optimal policy is precisely the $(1 - \beta)$-optimal policy with respect to the new process; and the optimal expected average cost is $\beta V'_{1-\beta}(0)$.

Thus, if our assumption holds, then we have reduced the average cost problem to a discounted cost problem and the methods of policy improvement or successive approximations may be employed.

GENERALIZATIONS. Theorem 6.17 easily generalizes to arbitrary state spaces. In Theorem 6.18, however, conditions are necessary to ensure that $h_\alpha(\cdot)$ has a convergent subsequence. In the case of a general state space, a sufficient condition (by the Ascoli Theorem) is that $h_\alpha(\cdot)$ be uniformly bounded and equicontinuous (see Ross [11]). The reduction to the discounted case follows in the same way when the state space is arbitrary. Our new assumption is, of course, that there exists some state, call it 0, and $\beta > 0$, such that

$$P\{X_{t+1} = 0 \mid X_t, a_t\} \geq \beta$$

for all states X_t and actions a_t.

6.8. Finite State Space—Computational Approaches

In this section we suppose that the state space, as well as the action space, is finite, and we denote the states by $0, 1, 2, \ldots, m$.

Discount Case

We shall first consider the policy improvement technique for the discount case. For any stationary policy f, we have by Remark 3 of Section 6.2 that V_f is the unique solution to

$$V_f(i) = C[i, f(i)] + \alpha \sum_{j=0}^{m} P_{ij}[f(i)]V_f(j), \qquad i = 0, 1, \ldots, m$$

Thus, we have $m + 1$ equations in $m + 1$ unknowns, and hence $V_f(i)$ may be obtained by standard methods. We may then improve upon f by choosing our actions so as to minimize

$$C(i, a) + \alpha \sum_{j=0}^{\infty} P_{ij}(a)V_f(j) \tag{29}$$

Let us also agree to only change our present action $f(i)$ if the new action leads to a *strict* improvement [over $f(i)$] in (29). Then it follows from Corollary 6.4 and Remark 2 of Section 6.2 that:

(i) If the improved policy is the original policy f, then f is α-optimal;
(ii) If the improved policy is not the original policy f, then the improved policy is strictly better than f for at least one initial state.

Once the improved policy is determined, its cost function may be calculated, and then it too may be improved. Since there are only a finite number of stationary policies, this policy improvement technique will eventually lead to an α-optimal policy.

Another computational approach to the discount problem follows from the following lemma.

Lemma 6.21

If T_α is defined as in (11), then for any function u,

$$T_\alpha u \geq u \Rightarrow V_\alpha \geq u$$

PROOF.   If $T_\alpha u \geq u$, then by the monotonicity of T_α it follows that $T_\alpha^n u \geq u$ and the result follows by letting $n \to \infty$.

Since $T_\alpha V_\alpha = V_\alpha$, it follows that V_α may be obtained by

Maximizing u
Subject to $T_\alpha u \geq u$

However, since maximizing $u(i)$ for each i will also maximize $\sum_{i=0}^m u(i)$, it follows that the problem reduces to

Maximizing $\sum_{i=0}^m u(i)$

Subject to $\min\left\{C(i, a) + \alpha \sum_{j=0}^m P_{ij}(a)u(j)\right\} \geq u(i), \qquad i = 0, 1, \ldots, m$

or equivalently to

Maximizing $\sum_{i=0}^m u(i)$

Subject to $C(i, a) + \alpha \sum_{j=0}^m P_{ij}(a)u(j) \geq u(i), \quad$ for all a, all $i = 0, 1, \ldots, m$

But this latter statement of the problem is just a linear program and may be solved by standard techniques of linear programming.

Average Cost Case—A Linear Programming Approach

We shall now consider the average cost problem, and we make the simplifying assumption that all stationary policies give rise to an irreducible Markov chain.

Rather than considering only stationary policies, we shall consider randomized stationary policies, and for a fixed policy of this sort, we let P_i^a denote the probability of taking action a when in state i. Also, let z_i, $i = 0, 1, \ldots, m$, be the vector of stationary probabilities (this exists, independent of the initial state, by our simplifying assumption). Then, letting

$$z_i^a = z_i P_i^a$$

it follows (see Problem 17) that the average cost is

$$\sum_i \sum_a z_i^a C(i, a) \tag{30}$$

subject to the restrictions

$$\sum_a z_i^a = \sum_j \sum_a z_j^a P_{ji}^a \tag{31}$$

$$z_i^a = z_i P_i^a \tag{32}$$

$$\sum_i \sum_a z_i^a = 1 \tag{33}$$

$$z_i^a \geq 0 \tag{34}$$

$$\sum_a z_i^a = z_i \tag{35}$$

The problem then reduces to a linear program of minimizing (30) subject to (31), (33) and (34). P_i^a then comes from (32) and (35).

It turns out, as it must by Corollary 6.20, that the minimal average cost can be achieved by a nonrandomized policy.

Problems

1. Consider a sequential decision process which is such that after each stage, there is a probability α that the process will end. Show that such a process falls under the discounted cost framework.

2. Let f_1 and f_2 be stationary policies and define f as follows:

$$f(i) = \begin{cases} f_1(i) & \text{if } V_{f_1}(i) \leq V_{f_2}(i) \\ f_2(i) & \text{if } V_{f_1}(i) > V_{f_2}(i) \end{cases}$$

Show that $V_f(i) \leq \min\{V_{f_1}(i), V_{f_2}(i)\}$, for all i.

3. Consider a machine that can be in any one of two states, *good* or *bad*. The machine produces items, which are either defective or nondefective, at the beginning of each day. The probability of a defective item is P_1 when in the good state and P_2 when in the bad state. Once in the bad state, the machine remains in this state until it is replaced. However, if the machine is in the good state at the beginning of one day, then with probability γ it will be in the bad state at the beginning of the next day. A decision upon whether or not to replace the machine must be made each day after observing the item produced. Let R be the cost of replacing the machine, and let C be the cost incurred whenever a defective item is produced. Set up the above as a Markov decision model and determine the functional equation satisfied by V_α. Assume that at time 0 there is a known probability that the machine is in the bad state.

4. Prove that for Problem 3, there is a P^* such that the α-optimal policy replaces whenever the present probability that the process is in the bad state is greater than or equal to P^*.

5. Prove Lemma 6.7.

6. In Example 2, show that $V_\alpha(P)$ is a concave, increasing function.

7. Consider the positive cost, no discounting case. Show that the policy improvement technique leads to a policy which is at least as good as the one improved. That is, for any policy f, let f^* be the policy which chooses the actions which minimize $C(i, a) + \sum_{j=0}^{\infty} P_{ij}(a)V_f(j)$. Show that $V_{f^*}(i) \leq V_f(i)$, all i.

8. Prove Lemma 6.11.

9. Consider a quiz show in which a contestant has n questions that he may attempt. The probability that he answers question i correctly is P_i, and the worth of question i is V_i, $i = 1, 2, \ldots, n$. The contestant must answer the questions one at a time, and once he misses a question he leaves the show (though he retains his earnings up to that point). If the contestant is allowed to answer questions in any order, what is the optimal ordering of the questions?

Hint: Let a policy be a permutation of $\{1, 2, \ldots, n\}$, and consider what happens when two successive numbers are interchanged. That is, compare the policy $(1, 2, 3, \ldots, n)$ with $(2, 1, 3, \ldots, n)$.

10. Consider a ball which may be in any one of n boxes. A search of the ith box costs $C_i > 0$ and finds the ball with probability α_i if the ball is in that box. Suppose that we are given prior probabilities P_i^0, $i = 1, 2, \ldots, n$, that the ball is in the ith box. Show that the policy which searches the box with the maximal value of $\alpha_i P_i/C_i$ minimizes the expected searching cost, where P_i is the posterior probability (given everything that has occurred up to that time) that the ball is in box i.

Hint: Define a policy to be a sequence $(\delta_1, \delta_2, \ldots)$, where $\delta_i \in (1, 2, \ldots, n)$ and δ_i represents the ith box to be searched. See what happens when two successive searches are interchanged.

11. To motivate the above search, suppose that a reward R_i is earned if the ball is found in the ith box. Suppose also that we may decide to stop at any time. Show how the results of Section 6.4 may be employed, and find the functional equation which determines the optimal policy.

12. In Problem 11, show that if $\sum \dfrac{C_i}{\alpha_i R_i} \leq 1$, then the optimal policy would never stop searching until the ball is found. Argue that if such is the case, then the policy given in Problem 10 is optimal.

13. Suppose that you flip a fair coin successively, with heads winning you one dollar and tails losing you one dollar. You may stop at any time. Show that this stopping rule problem is not stable.

14. We start a distance N parking places from our destination. As we cruise along, we can see only one parking place at a time. If a place is empty and we park there, then our loss is the distance we walk. Empty parking places occur independently with probability P. Solve this stopping problem by showing that it satisfies the conditions of Theorem 6.14.

15. Formulate the equivalent of Theorem 6.14 for the discount case.

16. Use Problem 15 to get an explicit solution for Example 3.

17. For a finite state space, show that the limit in (22) exists for any stationary policy, and also show that $\phi_f(i) = \sum_j z_j C[j, f(j)]$, where z_j is the limiting proportion of time spent in state j given that the initial state is i and f is used.

18. Do Problems 3 and 4 for the average cost criterion.

19. We say that the stationary policies f_n converge to the stationary policy f, and write $f_n \to f$, if for each i there exists an N_i such that $f_n(i) = f(i)$ for all $n \geq N_i$. Show that if $V_\alpha(i) - V_\alpha(0)$ is uniformly bounded in α and i, then for some sequence $\alpha_n \to 1$ and some average cost optimal policy f^*, $f_{\alpha_n} \to f^*$.

20. Suppose that $V_\alpha(i) - V_\alpha(0)$ is uniformly bounded. Show that if $f_{\alpha_n} \to f$ for some sequence $\alpha_n \to 1$, then f^* is average cost optimal.

21. Suppose that $P_{i0}(a) = \beta$ for all i, a. Show that for any stationary policy f, $\phi_f(0) = \beta V_f'(0, 1 - \beta)$, where $V_f'(0, 1 - \beta)$ is the expected $(1 - \beta)$-discount cost when the transition probabilities are given by (27).

Hint: Use the idea of a renewal reward process.

22. Argue that the result of Problem 21 is true whenever $P_{i0}(a) \geq \beta$ for all i, a.

23. If $V_\alpha(i) - V_\alpha(0)$ is uniformly bounded, show that

$$(1 - \alpha)V_\alpha(j) \to g \qquad \text{for all } j$$

24. Consider a Markov decision process in which a nonnegative reward $R(i, a)$ is earned when action a is taken in state i. Let $V_*(i)$ be the optimal (supremal) expected return function. Define the operator T mapping functions on the state space into itself by

$$(Tu)(i) = \max \{R(i, a) + \sum P_{ij}(a)u(j)\}$$

Show that $T^n 0 \to V_*$, and that $TV_* = V_*$. Use this to show that, if the state space is finite, the problem of obtaining V_* is equivalent to the following linear program:

$$\text{minimize} \sum_i u_i$$

subject to
$$u_i \geq R(i, a) + \sum_j P_{ij}(a)u_j \qquad \text{for all } i, \text{ all } a$$
$$u_i \geq 0, \qquad\qquad\qquad\qquad \text{all } i$$

References

The approach we have employed in the discounted cost case relies heavily on the results of Blackwell [1]. Example 1 is essentially due to Derman [3], who considered a finite state version. For a generalization of the countable version considered, the interested reader is referred to Ross [12]. The results of section 6.4 are from Strauch [13]. The approach to optimal stopping given in section 6.5 is new. However, with the exception of Theorem 6.13, the results of this section seem to be known. For a martingale approach to optimal stopping problems, the more advanced reader is referred to Chow-Robbins [2]. The sequential analysis Theorem 6.16 is due to Lehmann [8]. The first counterexample in Section 6.6 is due to Maitra (see [4]), and the second to Ross [10]. (See Derman [4] for a different example of the same phenomenon.) The results of Section 6.7 may be found in Ross [10] and [11]. For an analysis of the finite state case, the reader is referred to Howard [7]. The first person to have noticed that linear programming may be used to solve Markov decision models, seems to be Manne [9].

[1] BLACKWELL, D. "Discounted Dynamic Programming," *Annals of Mathematical Statistics*, **36**, pp. 226–235, (1965).

[2] CHOW, Y. S. and H. ROBBINS. "On Values associated with Stochastic Sequence," *Fifth Berkeley Symposium on Mathematical Statistics and Probability*. University of California Press, (1967).

[3] DERMAN, C. "On Optimal Replacement Rules when Changes of State are Markovian," *Mathematical Optimization Techniques*, R. Bellmen (ed.), (1963).

[4] DERMAN, C. "Denumerable State Markovian Decision Processes—Average Cost Criterion," *Annals of Mathematical Statistics*, **37**, pp. 1545–1554, (1966).

[5] DERMAN, C. and R. STRAUCH. "A Note on Memoryless Rules for Controlling Sequential Control Processes," *Annals of Mathematical Statistics*, **37**, pp. 276–279, (1966).

[6] FISHER, L. and S. ROSS. "An Example in Denumerable Decision Processes," *Annals of Mathematical Statistics*, **39**, pp. 674–675, (1968).

[7] HOWARD, R. "*Dynamic Programming and Markov Processes*," M.I.T. Press, Cambridge, Mass., (1960).

[8] LEHMANN, E. "*Testing Statistical Hypotheses*," Wiley Press, New York, 1959.

[9] MANNE, A. "Linear Programming and Sequential Decisions," *Management Science*, **6**, No. 3, pp. 259–267, (1960).

[10] ROSS, S., "Non-discounted Denumerable Markovian Decision Models," *Annals of Mathematical Statistics*, **39**, pp. 412–423, (1968).

[11] ROSS, S. "Arbitrary State Markovian Decision Processes," *Annals of Mathematical Statistics*, **39**, pp. 2118–2122, (1968).

[12] ROSS, S. "A Markovian Replacement Model with a Generalization to Include Stocking," *Management Science*, to appear (1969).

[13] STRAUCH, R. "Negative Dynamic Programming," *Annals of Mathematical Statistics*, **37**, pp. 871–890, (1966).

7

SEMI-MARKOV DECISION PROCESSES

7.1. Introduction

In this chapter we attempt to generalize the results of Chapter 6 to cover the case of a sequential decision process for which the times between transitions are random.

A process is observed at time 0 and classified into one of the states 0, 1, 2, ... After classification, one of a finite number of possible actions must be chosen. If the process is in state i and action a is chosen then:

(i) The next state of the process is chosen according to the transition probabilities $P_{ij}(a)$;

(ii) Conditional on the event that the next state is j, the time until the transition from i to j occurs is a random variable with probability distribution $F_{ij}(\cdot \mid a)$.

After the transition occurs, an action is again chosen and (i) and (ii) are repeated. This is assumed to go on indefinitely.

We further suppose that a cost structure is imposed on the model in the following manner: If action a is chosen when in state i, then an immediate cost $C(i, a)$ is incurred and, in addition, a cost rate $c(i, a)$ is imposed until the next transition occurs. That is, if a transition occurs after t units, then the total cost incurred is given by $C(i, a) + tc(i, a)$. Furthermore, we suppose that both $C(i, a)$ and $c(i, a)$ are bounded.

Clearly, when the transition times are identically one, then the above is just a Markov decision process, and in the general case, it is called a *semi-Markov decision process*. The reader should also note that if a stationary policy† is employed, then the process $\{X(t), t \geq 0\}$ is a semi-Markov process, where $X(t)$ represents the state of the process at time t.

In order to ensure that an infinite number of transitions does not occur in a finite interval, we shall assume throughout that the following conditions holds.

† For a definition of stationary policy see the previous chapter.

Condition 1

There exists $\delta > 0, \varepsilon > 0$, such that

$$\sum_{j=0}^{\infty} P_{ij}(a)F_{ij}(\delta \mid a) \leq 1 - \varepsilon \qquad \text{for all } i, a$$

Or, in other words, Condition 1 states that for every state i and action a, there is a positive probability of at least ε that the transition time will be greater than δ.

7.2. Discounted Cost Criterion

In this section we shall assume that costs are continuously discounted, and we shall attempt to minimize the expected total discounted cost. (By continuously discounting costs, we mean that a cost C incurred at time t is equivalent to a cost $Ce^{-\alpha t}$ at time 0.) For the most part, results follow in an identical fashion as in Section 6.2, and hence most of the theorems will be stated without proof. In places where definitions are not explicitly given, the reader is referred to Section 6.2.

Let X_n and a_n be respectively the nth state of the process and the nth action chosen, $n = 1, 2, \ldots$. Also, let $\tau_0 = 0$, and for $n > 0$ let τ_n be the time between the $(n - 1)$st and the nth transition. Now, for any policy π and $\alpha > 0$, let

$$V_\pi(i) = E_\pi \left[\sum_{n=1}^{\infty} e^{-\alpha(\tau_1 + \cdots + \tau_{n-1})} \left(C(X_n, a_n) + \int_0^{\tau_n} c(X_n, a_n)e^{-\alpha t} \, dt \right) \Bigg| X_1 = i \right]$$

$$(1)$$

Hence, $V_\pi(i)$ represents the expected total discounted cost incurred when the initial state is i and π is employed.

Also, letting

$$V_\alpha(i) = \inf_\pi V_\pi(i)$$

we say that π^* is α-optimal if

$$V_{\pi^*}(i) = V_\alpha(i), \qquad \text{all } i \geq 0$$

Theorem 7.1

$$V_\alpha(i) = \min_a \left\{ \bar{C}_\alpha(i, a) + \sum_{j=0}^{\infty} P_{ij}(a) \int_0^{\infty} e^{-\alpha t} V_\alpha(j) \, dF_{ij}(t \mid a) \right\} \qquad (2)$$

where

$$\bar{C}_\alpha(i, a) = C(i, a) + \sum_{j=0}^{\infty} P_{ij}(a) \int_0^{\infty} \int_0^t e^{-\alpha s} c(i, a) \, ds \, dF_{ij}(t \mid a)$$

is the expected one stage discounted cost incurred when action a is taken in state i.

If we now define the policy f_α to be the stationary policy which, when in state i, selects the (or an) action minimizing the right side of (2), we may prove—

Theorem 7.2

The stationary policy f_α is α-optimal. That is,

$$V_{f_\alpha}(i) = V_\alpha(i) \qquad \text{for all } i$$

REMARK. If for each stationary policy f, we define the mapping $T_f : B(I) \to B(I)$ by

$$(T_f u)(i) = \bar{C}_\alpha[i, f(i)] + \sum_{j=0}^{\infty} P_{ij}[f(i)] \int_0^\infty e^{-\alpha t} u(j) \, dF_{ij}[t \,|\, f(i)]$$

then by making use of Condition 1, the equivalent of Lemma 6.2 may be proven and then used to prove Theorem 7.2.

Let us now define the mapping $T_\alpha : B(I) \to B(I)$ by

$$(T_\alpha u)(i) = \min_a \left\{ \bar{C}_\alpha(i, a) + \sum_{j=0}^{\infty} P_{ij}(a) \int_0^\infty e^{-\alpha t} u(j) \, dF_{ij}(t \,|\, a) \right\}$$

We may now prove the following theorem.

Theorem 7.3

$\|T_\alpha u - T_\alpha v\| \leq (1 - \varepsilon + \varepsilon e^{-\alpha \delta}) \|u - v\|$ *for all $u, v \in B(I)$.†   That is, T_α is a contraction mapping with fixed point V_α; thus $T_\alpha^n u \to V_\alpha$ for all $u \in B(I)$.*

REMARKS AND GENERALIZATIONS. We leave it for the reader to prove the analogues of the remaining theorems of Section 6.2. Also, we note that our results, though stated for a countable state space, easily extend to noncountable state spaces. Furthermore, though we have assumed the specific cost structure of an immediate cost followed by a cost rate, this was done for notational purposes and is in no way essential. As a result, no matter how general the cost structure (as long as it follows certain boundedness conditions) Theorems 7.1–7.3 will remain true when $\bar{C}_\alpha(i, a)$ is interpreted as the expected one stage discounted cost incurred when action a is taken in state i.

† Where ε and δ are given by Condition 1.

7.3. Average Cost—Preliminaries and Equality of Criteria

In this section we give two possible definitions of the average expected cost, and then prove that under certain conditions they are equivalent.

Let $Z(t)$ denote the total cost incurred by time t, and let $Z_n = C(X_n, a_n) + \tau_n c(X_n, a_n)$ denote the cost incurred during the nth transition interval. Now, for any policy π, let

$$\phi_\pi^1(i) = \overline{\lim_{t \to \infty}} \, E_\pi\left[\frac{Z(t)}{t} \,\middle|\, X_1 = i\right]$$

and

$$\phi_\pi^2(i) = \overline{\lim_{n \to \infty}} \, \frac{E_\pi\left[\sum_{j=1}^n Z_j \,|\, X_1 = i\right]}{E_\pi\left[\sum_{j=1}^n \tau_j \,|\, X_1 = i\right]}$$

Thus, while ϕ^1 represents what we usually mean by the average expected cost, that is, the limit as $t \to \infty$ of the expected average cost incurred by time t, ϕ^2 also represents, at least in some sense, the average expected cost.

However, while ϕ^1 is clearly the more appealing criterion, it turns out that it is easier to work with ϕ^2. Fortunately, it turns out that under certain conditions, both criteria are equal. Roughly speaking, a sufficient condition is that for any stationary policy f, the resultant semi-Markov process $\{X(t), t \geq 0\}$ is a regenerative process with finite expected cycle length.

For any initial state i, let

$$T = \min\{t > 0 : X(t) = i, X(t^-) \neq i\}$$

and

$$N = \min\{n > 0 : X_{n+1} = i\}$$

Hence, T is the time of the first return to state i, and N is the number of transitions that it takes before this occurs. We shall need the following lemma.

Lemma 7.4

If $E_\pi[T \mid X_1 = i] < \infty$, then

$$E_\pi[N \mid X_1 = i] < \infty, \quad \text{and} \quad T = \sum_{n=1}^N \tau_n$$

PROOF.   By the definition of T and N, it follows that $T \geq \sum_{n=1}^N \tau_n$, with equality holding if $N < \infty$. Now, if we let

$$\bar{\tau}_n = \begin{cases} 0 & \text{if } \tau_n \le \delta \\ \delta \text{ with probability } \dfrac{\varepsilon}{1 - \sum_{j=0}^{\infty} P_{kj}(a)F_{kj}(\delta \mid a)} \\ & \qquad \text{if } \tau_n > \delta,\, X_n = k,\, a_n = a \\ 0 \text{ with probability } 1 - \dfrac{\varepsilon}{1 - \sum_{j=0}^{\infty} P_{kj}(a)F_{kj}(\delta \mid a)} \\ & \qquad \text{if } \tau_n > \delta,\, X_n = k,\, a_n = a \end{cases}$$

then it follows from Condition 1 that $\bar{\tau}_n$, $n = 1, 2, \ldots$ are independent and identically distributed with

$$P\{\bar{\tau}_n = \delta\} = \varepsilon = 1 - P\{\bar{\tau}_n = 0\}$$

Now, from Wald's equation, it follows that if $EN = \infty$ then $E \sum_{n=1}^{N} \bar{\tau}_n = \infty$, and hence that $ET \ge E \sum_{n=1}^{N} \tau_n \ge E \sum_{n=1}^{N} \bar{\tau}_n = \infty$ (since $\bar{\tau}_n \le \tau_n$). Therefore, if $ET < \infty$, then EN, and hence N, are finite, and the lemma is proven.

Theorem 7.5

If f is a stationary policy, and if $E_f[T \mid X_1 = i] < \infty$, then

$$\phi_f^1(i) = \phi_f^2(i) = \frac{E_f[Z(T) \mid X_1 = i]}{E_f[T \mid X_1 = i]}$$

PROOF. Suppose throughout the proof that $X_1 = i$. Now it is easily seen that, under a stationary policy, the process $\{X(t), t > 0\}$ is a regenerative process with regeneration point T. Hence, the process $\{Z(t), t \ge 0\}$ may be regarded as a renewal reward process, and thus by Theorem 3.16,

$$\phi_f^1(i) = \lim_{t \to \infty} \frac{E_f Z(t)}{t} = \frac{E_f Z(T)}{E_f T}$$

Now, it is also easy to see that $\{X_n, n = 1, 2, \ldots\}$ is a discrete time regenerative process with regeneration time N. Hence, by regarding $Z_1 + \cdots + Z_N$ as the *reward* earned during the first cycle of this process, it follows again by Theorem 3.16 that

$$E_f \sum_{i=1}^{n} \frac{Z_i}{n} \to \frac{E_f \sum_{i=1}^{N} Z_i}{E_f N} \qquad \text{as } n \to \infty \tag{3}$$

where we have used Lemma 7.4 in asserting that $E_f N < \infty$. However, we may also regard $\tau_1 + \cdots + \tau_N$ as the reward earned during the first cycle, and hence by the same reasoning,

$$E_f \sum_{i=1}^{n} \frac{\tau_i}{n} \to \frac{E_f \sum_{i=1}^{N} \tau_i}{E_f N} \qquad \text{as } n \to \infty \tag{4}$$

By combining (3) and (4), we obtain

$$\phi_f^2(i) = \frac{E_f \sum_{i=1}^{N} Z_i}{E_f \sum_{i=1}^{N} \tau_i} \tag{5}$$

However, since $N < \infty$ (Lemma 7.4), it is easy to see that $\sum_{i=1}^{N} Z_i = Z(T)$ and $\sum_{i=1}^{N} \tau_i = T$, and the result follows from (5).

REMARKS. It also follows from the above proof that, with probability 1,

$$\lim_{t \to \infty} \frac{Z(t)}{t} = \lim_{n \to \infty} \frac{\sum_{i=1}^{n} Z_i}{\sum_{i=1}^{n} \tau_i} = \frac{E_f Z(T)}{E_f T}$$

Also, suppose that the initial state is $j, j \neq i$. When is it true that $\phi_f^1(j) = \phi_f^2(j) = \phi_f^1(i)$? One answer is that if, with probability 1, the process will eventually enter state i, then $\{X(t), t \geq 0\}$ is a delayed regenerative process, and the proof goes through in an identical manner.

We now introduce some additional notation by letting

$$\bar{\tau}(i, a) = \sum_{j=0}^{\infty} P_{ij}(a) \int_0^{\infty} t \, dF_{ij}(t \mid a)$$

and

$$\bar{C}(i, a) = C(i, a) + c(i, a)\bar{\tau}(i, a)$$

In other words, $\bar{\tau}(i, a)$ is the expected time until a transition occurs when action a is taken in state i; and $\bar{C}(i, a)$ is the expected cost incurred during such a transition interval.

Let us now note that the expected cost incurred during a transition interval, and the expected length of a transition interval only, depend on the parameters of the process through $\bar{\tau}(i, a)$, $\bar{C}(i, a)$, and $P_{ij}(a)$; and, as a result, ϕ^2 only depends on the parameters of the process through the three functions $\bar{\tau}(i, a)$, $\bar{C}(i, a)$ and $P_{ij}(a)$. Hence, we may choose our cost and transition time distributions in as convenient a manner as possible. Namely, we may without loss of generality assume that $C(i, a) = \bar{C}(i, a)$, $c(i, a) = 0$, and that the time until transition when action a is chosen in state i is (with probability 1) $\bar{\tau}(i, a)$.

7.4. Average Cost—Results

In this section we shall attempt to parallel the development given in Section 6.7 for Markov decision processes. We start by proving the following generalization of Theorem 6.17.

Theorem 7.6

If there exists a bounded function $h(i)$, $i = 0, 1, 2, \ldots$, and a constant g such that

$$h(i) = \min_a \left\{ \bar{C}(i, a) + \sum_{j=0}^{\infty} P_{ij}(a)h(j) - g\bar{\tau}(i, a) \right\}, \qquad i \geq 0 \qquad (6)$$

then there exists a stationary π^* such that

$$g = \phi_{\pi^*}^2(i) = \min_\pi \phi_\pi^2(i), \qquad \text{for all } i \geq 0$$

and π^* is any policy which, for each i, prescribes an action which minimizes the right side of (6).

PROOF.   Let $H_n = (X_1, a_1, \ldots, X_n, a_n)$. For any policy π,

$$E_\pi \left\{ \sum_{i=2}^{n} [h(X_i) - E_\pi(h(X_i) \mid H_{i-1})] \right\} = 0$$

But,

$$E_\pi[h(X_i) \mid H_{i-1}] = \sum_{j=0}^{\infty} h(j)P_{X_{i-1}j}(a_{i-1})$$

$$= \bar{C}(X_{i-1}, a_{i-1}) + \sum_{j=0}^{\infty} h(j)P_{X_{i-1}j}(a_{i-1}) - g\bar{\tau}(X_{i-1}, a_{i-1})$$

$$- \bar{C}(X_{i-1}, a_{i-1}) + g\bar{\tau}(X_{i-1}, a_{i-1})$$

$$\geq \min_a \left\{ \bar{C}(X_{i-1}, a) + \sum_{j=0}^{\infty} h(j)P_{X_{i-1}j}(a) - g\bar{\tau}(X_{i-1}, a) \right\}$$

$$- \bar{C}(X_{i-1}, a_{i-1}) + g\bar{\tau}(X_{i-1}, a_{i-1})$$

$$= h(X_{i-1}) - \bar{C}(X_{i-1}, a_{i-1}) + g\bar{\tau}(X_{i-1}, a_{i-1})$$

with equality for π^* since π^* is defined to take the minimizing action. Hence,

$$0 \leq E_\pi \left\{ \sum_{i=2}^{n} [h(X_i) - h(X_{i-1}) + \bar{C}(X_{i-1}, a_{i-1}) - g\bar{\tau}(X_{i-1}, a_{i-1})] \right\},$$

or

$$g \leq \frac{E_\pi[h(X_n) - h(X_1)] + E_\pi \sum_{i=2}^{n} \bar{C}(X_{i-1}, a_{i-1})}{E_\pi \sum_{i=2}^{n} \bar{\tau}(X_{i-1}, a_{i-1})}$$

with equality for π^*. Now, letting $n \to \infty$ and using the boundedness of h and the fact that Condition 1 implies that $E_\pi \sum_{i=2}^{n} \bar{\tau}(X_{i-1}, a_{i-1}) \geq (n-1)\varepsilon\delta \to \infty$, we have that

$$g \leq \varlimsup_{n \to \infty} \frac{E_\pi \sum_{i=2}^n \bar{C}(X_{i-1}, a_{i-1})}{E_\pi \sum_{i=2}^n \bar{\tau}(X_{i-1}, a_{i-1})} = \phi_\pi^2(X_1)$$

with equality for π^* and for all values of X_1. Hence the desired result is proven.

In order to determine when the conditions of Theorem 7.6 are satisfied, we shall first turn to the discounted problem. Recall the fact that we are (without loss of generality) assuming that $C(i, a) = \bar{C}(i, a)$, $c(i, a) = 0$, and also that the transition time when action a is chosen when in state i is (with probability 1) $\bar{\tau}(i, a)$. Then we have by (2) that

$$V_\alpha(i) = \min_a \left\{ \bar{C}(i, a) + e^{-\alpha \bar{\tau}(i,a)} \sum_{j=0}^\infty P_{ij}(a) V_\alpha(j) \right\}, \qquad i \geq 0 \qquad (7)$$

Now, fix some state, say state 0, and define

$$h_\alpha(i) = V_\alpha(i) - V_\alpha(0)$$

Then, from (7), we obtain

$$h_\alpha(i) = \min_a \left\{ \bar{C}(i, a) + e^{-\alpha \bar{\tau}(i,a)} \sum_{j=0}^\infty P_{ij}(a) h_\alpha(j) + [e^{-\alpha \bar{\tau}(i,a)} - 1] V_\alpha(0) \right\}$$

$$= \min_a \left\{ \bar{C}(i, a) + e^{-\alpha \bar{\tau}(i,a)} \sum_{j=0}^\infty P_{ij}(a) h_\alpha(j) - V_\alpha(0)[\alpha \bar{\tau}(i, a) + o(\alpha)] \right\} \qquad (8)$$

Theorem 7.7

If $\bar{C}(i, a)$ *is bounded, and if there exists an* $N < \infty$ *such that*

$$|V_\alpha(i) - V_\alpha(0)| < N \qquad \text{for all } \alpha, \text{ all } i$$

then:

(i) There exists a bounded function $h(i)$ and a constant g satisfying (6);

(ii) For some sequence $\alpha_n \to 0$, $h(i) = \lim_{n \to \infty} (V_{\alpha_n}(i) - V_{\alpha_n}(0))$;

(iii) $\lim_{\alpha \to 0} \alpha V_\alpha(0) = g$.

PROOF.   By assumption, $h_\alpha(i)$ is uniformly bounded in α and i. Hence, since the state space is countable, we can, by Cauchy's diagonalization method, get a sequence $\alpha_n \to 0$ such that $\lim_{n \to \infty} h_{\alpha_n}(i) \equiv h(i)$ exists for all i. Also, since $\bar{C}(i, a)$ is bounded, it follows that $\alpha V_\alpha(0)$ is bounded, and hence we can require that $\lim_{n \to \infty} \alpha_n V_{\alpha_n}(0) \equiv g$ exists. The results (i) and (ii) then follow by letting $\alpha_n \to 0$ in (8). (Note that $o(\alpha_n) V_{\alpha_n}(0) = \alpha_n V_{\alpha_n}(0) o(\alpha_n)/\alpha_n \to 0$ as $n \to \infty$, since $\alpha_n V_{\alpha_n}(0)$ is bounded.)

The proof of (iii) follows in an identical manner as in Theorem 6.18.

7.5. Some Examples

EXAMPLE 1.   Suppose that letters arrive at a post office in accordance with a Poisson process with rate λ. At any time, the postmaster may, at a cost of K units, summon a truck to pick up all letters presently in the post office.† Suppose also that the post office incurs a cost at a rate $C(i)$ when there are i letters waiting to be picked up, where $C(i)$ is a bounded increasing non-negative function. The process is assumed to go on indefinitely, and the problem is to select a policy which minimizes the long-run average cost per unit time.

The preceding may be regarded as a two action semi-Markov decision process with states $1, 2, 3, \ldots$, where state i means that there are i letters waiting to be picked up. Action 1 is "summon a truck," and Action 2 is "don't summon a truck." (Note that since a truck would never be summoned if there were no letters in the post office, we need not have a state 0.)

The parameters of the process are:

$$P_{i1}(1) = 1, \ \bar{\tau}(i, 1) = 1/\lambda, \ \bar{C}(i, 1) = K + \frac{C(0)}{\lambda}, \qquad i \geq 1 \qquad (9)$$

$$P_{ii+1}(2) = 1, \ \bar{\tau}(i, 2) = 1/\lambda, \ \bar{C}(i, 2) = \frac{C(i)}{\lambda}, \qquad i \geq 1 \qquad (10)$$

This is so since if we summon a truck when in state i, then all items in the post office are immediately picked up, and the state will change (i.e., a transition will occur) only when the next letter arrives. Since the distribution of the time until the next letter arrives is exponential with mean $1/\lambda$, (9) follows. Similarly, if we do not dispatch, then the transition occurs when the next letter arrives, and hence (10) follows.

Now, if we let

$$V_\alpha(i, 1) = \min\left\{ K + \frac{C(0)}{\lambda}; \frac{C(i)}{\lambda} \right\}$$

and for $n > 1$,

$$V_\alpha(i, n) = \min\left\{ K + \frac{C(0)}{\lambda} + e^{-\alpha/\lambda}V_\alpha(1, n-1); \frac{C(i)}{\lambda} + e^{-\alpha/\lambda}V_\alpha(i+1, n-1) \right\}$$

then it is easily seen by induction that $V_\alpha(i, n)$ is increasing in i, and hence $V_\alpha(i) = \lim V_\alpha(i, n)$ is increasing in i. Also, since $V_\alpha(i)$ satisfies

$$V_\alpha(i) = \min\left\{ K + \frac{C(0)}{\lambda} + e^{-\alpha/\lambda}V_\alpha(1); \frac{C(i)}{\lambda} + e^{-\alpha/\lambda}V_\alpha(i+1) \right\}$$

† We assume that the truck arrives instantaneously.

it follows that

$$V_\alpha(i) \le K + \frac{C(0)}{\lambda} + e^{-\alpha/\lambda}V_\alpha(1)$$

$$< K + \frac{C(0)}{\lambda} + V_\alpha(1)$$

hence

$$V_\alpha(1) < V_\alpha(i) < V_\alpha(1) + K + \frac{C(0)}{\lambda}$$

Thus, by Theorem 7.7, it follows that there exists a constant g and a bounded increasing function $h(i)$ such that

$$h(i) = \min\left\{K + \frac{C(0)}{\lambda} + h(1) - \frac{g}{\lambda}; \frac{C(i)}{\lambda} + h(i + 1) - \frac{g}{\lambda}\right\}$$

and the policy which chooses the minimizing actions is optimal.

Now, if we let

$$i^* = \min\left\{i : \frac{C(i)}{\lambda} + h(i + 1) > K + \frac{C(0)}{\lambda} + h(1)\right\}$$

then it follows from the monotonicity of $C(i)$ and $h(i)$ that the optimal policy is to summon a truck whenever the number of letters in the post office is at least i^*.

However, though we have determined the structure of the optimal policy, it still remains to determine i^*. This may be done as follows:

Let f_i be the policy which summons a truck whenever there are at least i letters present, $i = 1, 2, \ldots$. Now, it is easily seen that each time the process enters state 1 it regenerates itself. Thus, if we say that a cycle is completed each time state 1 is reentered, it follows that the long-run average expected cost is given by

$$\phi_{f_i}(j) = \frac{E_{f_i}[\text{cost of cycle}]}{E_{f_i}[\text{length of cycle}]}$$

$$= \frac{K + \dfrac{C(0)}{\lambda} + E\displaystyle\int_{\tau_1}^{\tau_i} C[N(t)]\, dt}{i/\lambda}$$

where $N(t)$ is the total number of arrivals up to time t, and τ_n is the arrival time of the nth letter. Hence,

$$
\begin{aligned}
\phi_{f_i}(j) &= \frac{K + \dfrac{C(0)}{\lambda} + E[C(1)(\tau_2 - \tau_1) + \cdots + C(i-1)(\tau_i - \tau_{i-1})]}{i/\lambda} \\[2ex]
&= \frac{\lambda}{i}\left[K + \frac{C(0)}{\lambda} + \sum_{j=1}^{i-1} \frac{C(j)}{\lambda} \right] \\[2ex]
&= \frac{\lambda K}{i} + \frac{1}{i}\sum_{j=0}^{i-1} C(j)
\end{aligned}
$$

The optimal value of i, namely i^*, can then be found from (11).† For example, if $C(i) = iC$, then (11) equals $\lambda K/i + C(i-1)/2$, and by treating i as a continuous variable, we obtain by differential calculus that the optimal i is one of the two integers adjacent to $\sqrt{2\lambda K/C}$.

EXAMPLE 2. *The Streetwalker's Dilemma.* Consider a prostitute working in the city of San Francisco, and suppose that potential customers arrive in accordance with a Poisson process with rate λ. Each potential customer makes an offer consisting of the pair (i, F_i), where i is the amount of money offered and F_i is the distribution of the amount of time that the prostitute must spend with the customer. If the offer is rejected, then the arrival leaves and the prostitute waits for the next potential customer. If the offer is accepted, then all potential customers, who arrive while the prostitute is busy, are assumed lost. The successive offers are assumed independent and the offer (i, F_i) occurs with probability P_i, where $\sum_{i=1}^{N} P_i = 1$. The prostitute's dilemma is to choose her customers so as to maximize her long-run average return. (Those readers who do not like to think in the above terms may view this as a model for determining whether or not a factory should accept or reject potential jobs, when the factory is only able to handle one job at a time.)

The above may be viewed as a two action semi-Markov decision process with states $1, 2, \ldots, n$, where state i means that the prostitute must decide whether or not to accept an offer of (i, F_i). If we let $t_i = \int_0^\infty x \, dF_i(x)$ be the mean time spent with an i-type customer and let action 1 be the accept and action 2 the reject action, then the parameters of the process are given by

$$
\begin{aligned}
P_{ij}(1) &= P_j & P_{ij}(2) &= P_j \\
\bar{\tau}(i, 1) &= t_i + 1/\lambda & \bar{\tau}(i, 2) &= 1/\lambda \\
\bar{C}(i, 1) &= -i & \bar{C}(i, 2) &= 0
\end{aligned}
$$

---

† We are assuming that never dispatching is not optimal. Since the average cost in this case would be $\lim \sum_{i=1}^{n} C(i)/n$, it is easily seen when this is nonoptimal.

As in the previous example, we may suppose that the cost is immediately earned, and that the transition time is identically $\bar{\tau}(i, a)$. It is easy to check, via the discounted cost problem, that the conditions of Theorem 7.7 are satisfied, and hence by Theorem 7.6 the optimal average cost policy is the one which chooses actions so as to minimize the right side of

$$h(i) = \min\left\{ -i + \sum_{j=1}^{N} P_j h(j) - g(t_i + 1/\lambda); \quad \sum_{j=1}^{N} P_j h(j) - g/\lambda \right\}$$

Thus, the optimal policy accepts an offer (i, F_i) if and only if

$$\frac{i}{t_i} \geq -g$$

where $-g$ is the optimal average return per time. Hence, the structure of the optimal policy is determined.

Problems

1. Show that (1) is bounded.
2. Prove Theorem 7.1.
3. Prove Theorem 7.2.
4. Prove Theorem 7.3.
5. Show that the policy improvement technique remains valid in the semi-Markov case.
6. State and prove the analog of Corollary 6.6.
7. Formulate and prove the analog to Theorem 6.19.
8. Solve the example of Section 5 when there is a lag of L time units between the summoning of the truck and the actual arrival of the truck. Let L be random.
9. Solve the example of Section 5 when the criterion is the total expected discounted cost.
10. Suppose that for the example in Section 5, arrivals consist not of single letters but of batches of letters. Suppose that the batches are independent and have some common distribution. (Hence the arrival stream of letters is a compound Poisson process.) Determine the structure of the optimal policy.
11. Consider an $M/G/1$ queueing system in which the server may be either on or off. Suppose that costs are incurred at a rate C_1 when the server is on and C_2 when he is off. Assume a holding cost rate of h per customer. Finally, suppose that a fixed cost of A_1 is incurred when the server goes from on to off, and A_2 when he goes from off to on. The server may switch from off to on at any time. However, he may switch from on to off only if he is not presently serving a customer. Formulate the above as a semi-Markov decision process and solve.
12. Consider the equivalent problem for $G/M/1$ queueing system and solve. Assume that the server can only switch at arrival points of customers.
13. Consider a machine which can be in one of the states $0, 1, 2, \ldots$. The state of the machine is initially noted and a decision upon whether or not to replace it is made.

If the decision to replace is made, then the machine is instantaneously replaced by a new machine whose state is 0. If the machine is in state i and it is not replaced, then it remains in state i for a random amount of time and then enters some other state. Assume that there is a cost R incurred whenever a replacement is made and also that a cost is incurred at a rate $c(i)$ whenever the machine in use is in state i.

Formulate the above as a semi-Markov decision process. What is the state space? Is there a state 0? Why not? Impose conditions on the parameters of the process which imply that the optimal policy has a nice form.

14. Solve the streetwalker's dilemma when offers come in batches. The streetwalker is allowed to accept at most one offer.

References

The results of this chapter are taken from a recent paper of S. Ross [4]. This paper also allows for uncountable state spaces. The second example is taken from a paper of Lippman and Ross [3].

[1] Howard, R., "Semi-Markovian Decision Processes," *International Statistical Institute Bulletin*, (1963).

[2] Jewell, W., "Markov Renewal Programming I and II," *Operations Research*, **2**, No. 6, pp. 938–971, (1963).

[3] Lippman, S. and S. Ross, "The Streetwalker's Dilemma: A Job Shop Model," Western Management Science Institute Technical Report, University of California, Los Angeles, (November 1969).

[4] Ross, S., "Average Cost Semi-Markov Decision Processes," ORC 69-27, Operations Research Center, University of California, Berkeley, (September 1969).

8

INVENTORY THEORY

8.1. Introduction

In previous chapters, we have made a probabilistic analysis of inventory systems in which (s, S) policies were employed. For instance, by viewing the costs incurred as a renewal reward process, we have shown how to calculate the long-run average cost connected with an (s, S) policy (see Problem 18 of Chapter 3); and by viewing the changing inventory level as a regenerative process, we have shown how to calculate the limiting distribution of the inventory level (see Section 5.4).

In this chapter, we shall be concerned with the optimization problem. That is, rather than analyze (s, S) policies, we shall consider the question of whether or not we are justified in only considering this class of policies. This and other optimality questions relating to inventory models will be discussed in this chapter.

8.2. A Single Period Model

The following inventory model is to be considered. Items are produced (or purchased) for a single period. The cost of producing z units is

$$C(z) = \begin{cases} K + c \cdot z & \text{if } z > 0 \\ 0 & \text{if } z = 0 \end{cases}$$

where $K > 0$. In words, there is a cost of c dollars per item, and in addition there is a setup cost K incurred whenever a purchase is made.

Further, we suppose that there is a holding cost of h dollars for each item remaining at the end of the period, and a shortage cost of p dollars for each unit of unmet demand. The problem is: Given a probability density $f(\xi)$ of the one period demand, and an initial inventory x, then, under the assumption that the order is immediately filled, how much additional inventory should be ordered so as to minimize our total expected cost?

If we let

$$L(y) = p \int_y^\infty (\xi - y)f(\xi)\,d\xi + h \int_0^y (y - \xi)f(\xi)\,d\xi \tag{1}$$

then $L(y)$ represents the expected penalty and holding cost incurred if we order so as to bring our inventory up to y, i.e., if we order $y - x$. Hence, the total expected cost if we order up to y is given by

$$\begin{cases} K + c \cdot (y - x) + L(y) & \text{if } y > x \\ L(x) & \text{if } y = x \end{cases} \tag{2}$$

It should be noted that it easily follows from (1) that $L(y)$ is a strictly convex function.

Now, let us define S to be the value of y which minimizes $cy + L(y)$, and define s to be the smallest value of y for which $cs + L(s) = K + cS + L(S)$ (see Figure 8.1). From Figure 8.1 [or, more accurately, from the convexity

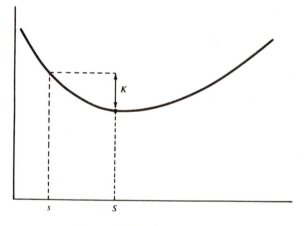

Figure 8.1 Graph of $cy + L(y)$

of $cy + L(y)$], it is evident that if $x > S$, then $cy + L(y) > cx + L(x)$ for all $y > x$; and thus $K + c(y - x) + L(y) > L(x)$ for all $y > x$. Hence, the optimal policy does not order if $x > S$. If $s \le x \le S$, then it is evident from Figure 8.1 that

$$K + cy + L(y) \ge cx + L(x) \qquad \text{for all} \quad y > x$$

and thus

$$K + c(y - x) + L(y) \ge L(x) \qquad \text{for all} \quad y > x$$

Again, no ordering is less costly than ordering. Finally, if $x < s$, then from Figure 8.1,

$$\min_{y \geq x}[K + cy + L(y)] = K + cS + L(S) < cx + L(x)$$

or

$$\min_{y \geq x}[K + c(y - x) + L(y)] = K + c(S - x) + L(S) < L(x)$$

Thus, the optimal policy orders up to S.
As a result, the optimal policy is

$$\begin{cases} \text{if} \quad x < s, & \text{order up to } S \\ \text{if} \quad x \geq s, & \text{do not order} \end{cases}$$

The value of S (and hence of s) may then be derived by minimizing $G(y) = cy + L(y)$. Upon differentiation, this yields that S is that value such that

$$F(S) = \frac{p - c}{p + h}$$

where F is the cumulative distribution function of the demand.

Note also that if there is no setup cost, i.e., if $K = 0$, then $S = s$, and the optimal policy is

$$\begin{cases} \text{if} \quad x < s & \text{order up to } s \\ \text{if} \quad x \geq s & \text{do not order} \end{cases}$$

8.3. Multi-Period Models

We shall now consider the n-period version of the model of Section 8.2. We suppose that the demands for the successive periods are independent and identically distributed. We again assume that the purchase cost of z items is of the form

$$C(z) = \begin{cases} K + cz & \text{if} \quad z > 0 \\ 0 & \text{if} \quad z = 0 \end{cases}$$

and that there is a holding cost of h dollars for each item of remaining inventory at the end of a period and a shortage cost of p per item whenever demand exceeds supply. It is also assumed that if the demand exceeds the supply, then the additional demand is backlogged and is filled when additional inventory becomes available (this backlogging is represented as negative inventory). Finally, we assume that all items remaining at the end of the last period have 0 value, and that all unfilled demand at this time is lost. The problem is to minimize our total expected cost incurred over the n periods, under the assumption that our orders are instantly filled.

If the stock level immediately after purchase is y, then the expected shortage and holding cost for that period is

$$L(y) = \begin{cases} h \int_0^y (y - \xi)f(\xi)\,d\xi + p \int_y^\infty (\xi - y)f(\xi)\,d\xi & y \geq 0 \\ p \int_0^\infty (\xi - y)f(\xi)\,d\xi & y < 0 \end{cases}$$

where f is the demand density. For the linear cost structure we have assumed $L(y)$ is easily seen to be convex.

Let us assume a discount factor α, $0 \leq \alpha \leq 1$, and let $V_n(x)$ be the minimal expected discounted cost incurred through the first n periods. It is easy to see that $V_n(x)$ satisfies

$$V_0(x) = 0$$

and

$$V_n(x) = \min_{y \geq x} \left\{ K\delta(y, x) + c(y - x) + L(y) + \alpha \int_0^\infty V_{n-1}(y - \xi)f(\xi)\,d\xi \right\} \quad (3)$$

where

$$\delta(y, x) = \begin{cases} 1 & y > x \\ 0 & y = x \end{cases}$$

We shall show that the optimal policy for this n period problem is given by the sequence of critical numbers (s_1, S_1), (s_2, S_2), ..., (s_n, S_n), along with the instruction that if there are j periods to go and our inventory is x, then we should order

$$\begin{cases} S_j - x & \text{if } x < s_j \\ 0 & \text{if } x \geq s_j \end{cases}$$

This result will be obtained by a study of the functions

$$G_n(y) = cy + L(y) + \alpha \int_0^\infty V_{n-1}(y - \xi)f(\xi)\,d\xi \quad (4)$$

Note that by (3) it is optimal to order at the initial period (of an n period problem) when our inventory is x, if there is a $y > x$ such that $G_n(x) > K + G_n(y)$; if we do order from x, then it will be to that $y > x$ which minimizes $G_n(y)$.

If $G_n(y)$ were convex, then it would follow, exactly as in the previous section, that an (s_j, S_j), $j = 1, \ldots, n$, policy is optimal. Unfortunately, it can be shown by numerical calculation that $G_n(y)$ need not be convex. However, what we can and will show is the following:

$$K + G_n(a + x) - G_n(x) - aG_n'(x) \geq 0 \quad \text{for all } a \geq 0, \text{ all } x \quad (5)$$

and (5) implies that the optimal policy is (s_j, S_j), $j = 1, \ldots, n$, where S_j is defined to minimize $G_j(y)$, and $s_j < S_j$ is defined by $G_j(S_j) = K + G_j(s_j)$.

To see why (5) implies that the optimal policy is (s_j, S_j), let us examine a graph of $G_n(x)$, which illustrates a typical case in which more complex policies are to be expected (see Figure 8.2). With this type of graph for G_n,

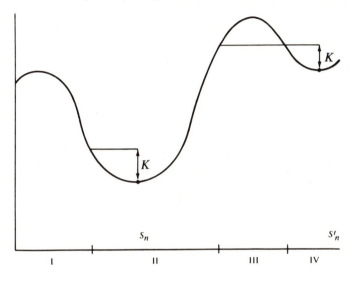

Figure 8.2.

we would order in interval I to S_n, not order in interval II, order in interval III to S'_n, and not order in IV. But if (5) is correct, this sort of graph is impossible. To see this, let x_0 be the point in I at which the relative maximum is attained, and let $a = S_n - x_0$. Then, $G'_n(x_0) = 0$ by the calculus, and (5) implies that

$$K + G_n(S) - G_n(x_0) \geq 0$$

which contradicts the graph. The same argument may be applied to the point S'_n. Hence, (s_n, S_n) is optimal for the initial period of an n period model, and the result follows.

Before proving that (5) holds, it will be convenient to define a concept known as *K-convexity*.

Definition

For any number $K \geq 0$, the differentiable function $g(x)$ is said to be *K*-convex if

$$K + g(a + x) - g(x) - ag'(x) \geq 0 \qquad \text{for all} \quad a > 0, \text{ all } x$$

Hence, (5) asserts that G_n is K-convex. The following properties of K-convex functions are easily proved and are left as an exercise:

(i) 0-convexity is equivalent to ordinary convexity.

(ii) If $g(x)$ is K-convex, then $g(x + h)$ is K-convex for all h.

(iii) If $g_1(x)$ and $g_2(x)$ are K-convex and N-convex respectively, then $\alpha g_1(x) + \beta g_2(x)$ is $(\alpha K + \beta N)$-convex, for all positive α and β.

(iv) If $g(x)$ is K-convex and f is a probability density, then

$$\alpha \int_0^\infty g(y - \xi) f(\xi)\, d\xi \qquad \text{is } K\text{-convex for all} \quad 0 < \alpha \le 1$$

We are now ready to prove (5), namely, that $G_n(y)$ is K-convex. This will be done inductively. Now, $G_1(y) = cy + L(y)$ is 0-convex [since $L(y)$ is convex] and hence K-convex. So assume that $G_1, \ldots, G_n$ are K-convex. By (4) and Properties (iii) and (iv), it will follow that G_{n+1} is K-convex if we can show that V_n is K-convex.

The K-convexity of V_n is shown as follows. Since, by assumption, G_n is K-convex, it follows from our previous argument that the optimal policy for the initial period of an n period problem is (s_n, S_n), and hence from (3) and (4),

$$V_n(x) = \begin{cases} K + G_n(S_n) - cx & x < s_n \\ G_n(x) - cx & x \ge s_n \end{cases} \tag{6}$$

We will now use (6) to prove that V_n is K-convex. So, choose an $a > 0$. We must consider three cases.

CASE 1. $x \ge s_n$. In this region, $V_n(x)$ is the sum of linear function and a K-convex function, and is thus K-convex.

CASE 2. $x < s_n < x + a$. In this case,

$$K + V_n(x + a) - V_n(x) - a V_n'(x) = K + V_n(x + a) - V_n(x) + ac$$

and this is positive, since

$$V_n(x) = \min_{y > x} \left\{ K + c \cdot (y - x) + L(y) + \alpha \int_0^\infty V_{n-1}(y - \xi) f(\xi)\, d\xi \right\}$$

$$\le K + ca + L(x + a) + \alpha \int_0^\infty V_{n-1}(x + a - \xi) f(\xi)\, d\xi$$

$$= K + ca + V_n(x + a) \tag{7}$$

[Note that (7) follows, since we must order at x as $x < s_n$; and the final equation follows, since we do not order at $x + a$ as $x + a > s_n$.]

CASE 3. $x + a < s_n$.   In this region, $V_n(x)$ is linear and hence K-convex. This completes the induction, and demonstrates the optimality of an (s_j, S_j), $j = 1, \ldots, n$ policy.

Finally, if we let the discount factor α be strictly less than 1, and consider the infinite period problem, then by letting $n \rightarrow \infty$ it is possible to demonstrate that an (s, S) policy is optimal.

8.4. A Multi-Period Stationary Optimal Policy

Consider now an n period model similar to the one just presented, but which assumes that at the end of the n periods, (i) all unfilled demand must be satisfied by purchasing items at a cost of c per unit, and (ii) if any stock is left over, then it may be returned at cost price. Furthermore, we suppose that there is no setup cost, i.e., $K = 0$, and we let the discount factor be unity.

Again, let $V_n(x)$ be the minimal expected cost for an n period problem given that the initial inventory is x. Then,

$$V_1(x) = \min_{y \geq x} \left\{ c \cdot (y - x) - c \int_0^\infty (y - \xi) f(\xi) \, d\xi + L(y) \right\}$$

$$= \min_{y \geq x} L(y) + c\mu - cx$$

where $\mu = \int_0^\infty \xi f(\xi) d\xi$ is the expected demand. Hence, letting s be the minimizing value of $L(y)$ (s is such that $F(s) = p/p + h$) it follows, from the convexity of $L(y)$, that the optimal one period problem orders to s if $x \leq s$, and does not order if $x > s$. Hence,

$$V_1(x) = \begin{cases} L(s) + c\mu - cx & x \leq s \\ L(x) + c\mu - cx & x > s \end{cases} \tag{8}$$

Now, let us consider the two period problem. It follows that

$$V_2(x) = \min_{y \geq x} \left\{ c \cdot (y - x) + \int_0^\infty V_1(y - \xi) f(\xi) \, d\xi + L(y) \right\} \tag{9}$$

Now, when $y = s$,

$$cy + L(y) + \int_0^\infty V_1(y - \xi) f(\xi) \, d\xi = 2[L(s) + c\mu]$$

Also, since $L(y)$ is minimized at $y = s$, we have from (8) that $V_1(y) \geq L(s) + c\mu - cy$; hence,

$$cy + L(y) + \int_0^\infty V_1(y - \xi) f(\xi) \, d\xi \geq 2c\mu + L(y) + L(s) \geq 2[L(s) + c\mu]$$

Thus, $G_2(y) = cy + L(y) + \int_0^\infty V_1(y - \xi) f(\xi) \, d\xi$ is minimized at $y = s$. Now, it follows from (8) that V_1 is convex and thus $G_2(y)$ is convex. From this fact and (9), it follows that the optimal policy for the two stage problem orders up to s if the initial inventory x is less than s, and does not order otherwise. Hence, there is a single critical number for the first two periods.

Furthermore, we have shown that

$$V_2(x) = \begin{cases} G_2(s) - cx = 2[L(s) + c\mu] - cx & x \le s \\ G_2(x) - cx \ge 2[L(s) + c\mu] - cx & x > s \end{cases}$$

and hence $V_2(x)$ is convex [since $G_2(x)$ is].

In fact, we may show by induction that

$$V_n(x) = \begin{cases} G_n(s) - cx = n[L(s) + c\mu] - cx & x \le s \\ G_n(x) - cx \ge n[L(s) + c\mu] - cx & x > s \end{cases} \tag{10}$$

where

$$G_n(x) = cx + L(x) + \int_0^\infty V_{n-1}(y - \xi)f(\xi)\,d\xi \tag{11}$$

and where both $V_n(x)$ and $G_n(x)$ are convex. Since

$$V_n(x) = \min_{y \ge x} G_n(y) - cx$$

this enables us to conclude that the optimal policy for an n period model is to order up to s whenever the inventory level is less than s, and not to order otherwise. Hence, a single critical number suffices for all periods.

8.5. Inventory Issuing Policies

A different sort of inventory model from the ones previously considered is the following. A stockpile consists of n items. Associated with the ith item is an initial age x_i, $i = 1, \ldots, n$. The field life of the item is a function, $L(x)$, of the age of the item upon issue to the field. When an item's life in the field is ended, a replacement is issued from the stockpile. Items are to be issued successively until the stockpile is depleted. The problem of interest is to find the order of item issue which maximizes the total field life obtained from the stockpile.

Of the $n!$ possible issuing policies, there are two which immediately come to mind. The first policy, called *LIFO* (*last in, first out*), issues the items in increasing order of age, while the second policy, called *FIFO* (*first in, first out*), issues the items in decreasing order of age.

As an example, suppose the stockpile consists of two items, with respective ages x_1 and x_2, where $x_1 > x_2$. Then, the total field life following a *LIFO* policy is given by

$$L(x_2) + L[x_1 + L(x_2)] \tag{12}$$

while the total field life following a *FIFO* policy is given by

$$L(x_1) + L[x_2 + L(x_1)] \tag{13}$$

We shall now attempt to determine conditions under which either the *LIFO* or *FIFO* issuing policy is optimal. We will consider first conditions for *LIFO* and then for *FIFO*.

Proposition 8.1

If

(i) $\dfrac{dL(x)}{dx} = L'(x) \geq -1$

and if

(ii) *LIFO* is an optimal policy when $n = 2$ [that is, if (12) > (13) whenever $x_1 > x_2$],

then *LIFO* is an optimal policy for all $n \geq 2$.

PROOF. The proof is by induction, and so assume that *LIFO* is optimal whenever there are $n - 1$ items to be issued. Now, for the n item case, let x be the initial age of the last item used, and let T be the total field life of the first $(n - 1)$ items issued, when an optimal policy is employed. Hence, the total field life will be

$$T + L(x + T)$$

which is increasing in T (since $L' \geq -1$). Therefore, for fixed x, $T + L(x + T)$ will attain its maximal value when T is as large as possible, and thus by the induction hypothesis, the first $n - 1$ items used should be ordered according to a *LIFO* policy. If the last item to be issued is the oldest, then *LIFO* is optimal and the induction is complete. However, this must be the case, for if the last item issued were not the oldest, then the oldest would be issued next to last (since the first $n - 1$ are ordered by *LIFO*); hence we could increase our field life by switching the last two (since *LIFO* is optimal for $n = 2$). The proposition is thus proven.

The corresponding proposition for the optimality of *FIFO* ordering will now be stated. Its proof is identical with the one just given and hence will not be presented.

Proposition 8.2

If

(i) $\dfrac{dL}{dx} \geq -1$

and if
(ii) *FIFO* is an optimal policy when $n = 2$,

then *FIFO* is an optimal policy for all n.

Summarizing the previous propositions yields the following assertion. If $L'(x) \geq -1$, and if *LIFO* (*FIFO*) is optimal for $n = 2$, then it is optimal for all n. Further sufficient conditions for the optimality of *LIFO* and *FIFO* are given as exercises.

The inventory issuing model just considered is a special case of the following model. Consider a stockpile consisting of n items, the ith item initially having a rating r_i, $i = 1, 2, \ldots, n$. We suppose that the rating of an item changes in time, and we let $d(r, t)$ denote the rating at time t of an item whose rating was r at time 0. If an item is placed into the field when its rating is r, then its field life is given by some function $L(r)$. The problem, as before, is to issue the items so as to maximize the total field life. The previous model is thus obtained when $d(r, t) = r + t$.

Another interesting model is obtained when $d(r, t) = re^{-\alpha t}$. If we assume that $L(r)$ is a nondecreasing function, then when $\alpha \geq 0$, we have the case of exponential deterioration. On the other hand, if $L(r)$ is nondecreasing and $\alpha < 0$, then we may suppose that the items are being improved while they remain in the stockpile. The following result may be proven.

Proposition 8.3

If $L(r) = r$ and $d(r, t) = re^{-\alpha t}$, then *FIFO* is an optimal issuing policy for all n.

Problems

1. For the model of Section 3, show directly that $G_n(y)$ is convex when $K = 0$. Argue from this that the optimal policy is determined by a single critical number.

2. Prove the four properties of K-convexity which are quoted in Section 3.

3. Prove that if $L(x)$ is a nonincreasing convex function, and if *LIFO* is an optimal policy when $n = 2$, then *LIFO* is optimal for all n.

4. Prove that if $L(x)$ is a nonincreasing or nondecreasing concave function which is such that $L'(x) \geq -1$, then *FIFO* is optimal for all n.

5. Consider the inventory issue age model of Proposition 8.3, but suppose that when an item of age x is put into the field, its lifetime is an exponential random variable with mean x. Show that *FIFO* is optimal when $n = 2$. It is not known if *FIFO* is optimal for all n.

6. Prove Proposition 8.3.

References

The notion of K-covexity and the ingenious proof of Section 3 are due to Scarf [3]. The results of Section 4 are due to Veinott [4]. Propositions 1 and 2 of Section 5 are due to Lieberman [2].

[1] ARROW, K. J., S. KARLIN and H. SCARF, *Studies in the Mathematical Theory of Inventory and Production*, Stanford University Press, Stanford, California, (1958).

[2] LIEBERMAN, G. J., "*LIFO* Vs. *FIFO* in Inventory Depletion Management," *Management Science*, **5**, No. 2, pp. 102–105, (1958).

[3] SCARF, H., "The Optimality of (s, S) Policies for the Dynamic Inventory Problem," *Proceedings of the First Stanford Symposium on Mathematical Methods in the Social Sciences*, Stanford University Press, Stanford, California, (1960).

[4] VEINOTT, A. F., Jr., "The Optimal Inventory Policy for Batch Ordering," *Operations Research*, **13**, No. 3, pp. 424–432, (1965).

[5] VEINOTT, A. F., Jr., "The Status of Mathematical Inventory Theory," *Management Science*, **12**, No. 11, pp. 745–777, (1966).

9

BROWNIAN MOTION AND CONTINUOUS TIME OPTIMIZATION MODELS

9.1. Introduction and Preliminaries

One of the most useful stochastic processes in applied probability theory is the Wiener process. The Wiener process originated in physics as a description of Brownian motion. This phenomenon, named after the English botanist Robert Brown who discovered it, is the motion exhibited by a small particle which is totally immersed in a liquid or gas. Since then, the Wiener process has been used beneficially in such areas as statistical testing of goodness of fit, analyzing the price levels on the stock market and quantum mechanics.

The first explanation of the phenomenon of Brownian motion was given by Einstein in 1905. He showed that Brownian motion could be explained by assuming that the immersed particle was continually being subject to bombardment by the molecules of the surrounding medium. However, the first concise definition of this stochastic process underlying Brownian motion was given by Wiener in a series of papers originating in 1918. The definition of this process, known as the Wiener process, or simply the *Brownian motion process*, is as follows.

A stochastic process $[X(t), t \geq 0]$ is said to be a Wiener process with drift coefficient μ if:

(i) $X(0) = 0$;

(ii) $\{X(t), t \geq 0\}$ has stationary independent increments;

(iii) for every $t > 0$, $X(t)$ is normally distributed with mean μt.

By writing

$$X(t + s) = X(t + s) - X(t) + X(t)$$

and using the assumption of independent increments, we obtain

$$\text{Var}[X(t + s)] = \text{Var}[X(t + s) - X(t)] + \text{Var}[X(t)]$$

which by stationary increments, yields

$$\text{Var}[X(t + s)] = \text{Var}[X(s)] + \text{Var}[X(t)]$$

However, the solution of the above functional equation is just

$$\text{Var}[X(t)] = \sigma^2 t$$

The value of σ^2 is a function of the underlying process and must be empirically determined. We shall assume throughout that $\sigma^2 = 1$, and hence the probability density of $X(t)$ is given by

$$f_t(x) = \frac{1}{\sqrt{2\pi t}} \exp\left[\frac{-(x - \mu t)^2}{2t}\right]$$

From the stationary independent increment assumption, it easily follows that the joint distribution of $X(t_1), \ldots, X(t_n)$ is given by

$$f_{t_1,\ldots,t_n}(x_1, \ldots, x_n) = f_{t_1}(x_1)f_{t_2-t_1}(x_2 - x_1) \cdot \ldots \cdot f_{t_n-t_{n-1}}(x_n - x_{n-1}) \quad (1)$$

By using (1), we may compute in principle any conditional probabilities desired. For instance, it is easy to show that the conditional distribution of $X(t)$ given that $X(t_1) = A$ and $X(t_2) = B$, when $t_1 < t < t_2$, is just normal with mean

$$A + \mu(t - t_1) + \frac{[B - A + \mu(t_1 - t_2)]}{t_2 - t_1}(t - t_1) \quad (2)$$

and variance

$$\frac{(t_2 - t)(t - t_1)}{t_2 - t_1} \quad (3)$$

The physical origins of the Wiener process suggest that the sample path should be a continuous function of t. This turns out to be the case, and it may be proven that, with probability 1, $X(t)$ is indeed a continuous function. This fact is quite deep, and no proof shall be attempted. Also, we should note in passing that while the sample path $X(t)$ is continuous, it is in no way ordinary or run-of-the-mill. As a matter of fact, it turns out that, with probability 1, $X(t)$ is nowhere differentiable.

9.2. Maximum of the Wiener Process

In this section we assume that the drift coefficient μ is zero, and we make use of the path continuity of the Wiener process in order to determine the distribution of $\max_{0 \le t \le T} X(t)$. This is accomplished by first noting that the process is symmetric about the x axis (since $\mu = 0$), and then exploiting this fact via the reflection principle.

Consider any sample path $X(t)$ for which $X(T) \geq a$, and let τ denote the first time that this path hits a. For $t > \tau$, let us reflect $X(t)$ about the line $x = a$ to obtain

$$\tilde{X}(t) = \begin{cases} X(t) & t < \tau \\ a - [X(t) - a] & t \geq \tau \end{cases}$$

Note that $\max_{0 \leq t \leq T} X(t) \geq a$ and $\max_{0 \leq t \leq T} \tilde{X}(t) \geq a$, and furthermore that, by symmetry, $X(t)$ and $\tilde{X}(t)$ have the same probability law. Hence, for every sample path $X(t)$ for which $X(T) \geq a$, there correspond two paths for which $\max_{0 \leq t \leq T} X(t) \geq a$. Furthermore, the converse of this is true: namely, that every sample for which $\max_{0 \leq t \leq T} X(t) \geq a$ corresponds to either of two sample paths $X(t)$ with equal probability, one of which is such that $X(T) > a$ [unless $X(T) = a$, which fortunately has 0 probability]. Hence,

$$P\left\{ \max_{0 \leq t \leq T} X(t) \geq a \right\} = 2P\{X(T) \geq a\}$$

$$= \frac{2}{\sqrt{2\pi T}} \int_a^\infty \left[\exp\left(\frac{-x^2}{2t} \right) \right] dx \tag{4}$$

If we let τ_a denote the time at which a Wiener process (with zero drift) reaches a, then from (4),

$$P\{\tau_a \leq t\} = P\left\{ \max_{0 \leq s \leq t} X(s) \geq a \right\}$$

$$= \frac{2}{\sqrt{2\pi t}} \int_a^\infty \left[\exp\left(\frac{-x^2}{2t} \right) \right] dx$$

$$= \frac{2}{\sqrt{2\pi}} \int_{a/\sqrt{t}}^\infty \left[\exp\left(\frac{-y^2}{2} \right) \right] dy. \tag{5}$$

Note that from (5), it follows that τ_a is finite with probability 1.

9.3. The Wiener Process and Optimization

EXAMPLE 1.   *Exercising a Stock Option.*   The first example we deal with considers the question of when to exercise an option in the stock market. Specifically, suppose that we have the option of buying, at some time in the future, one unit of a stock at a fixed price A, independent of its current market price. The current market price of the stock is taken to be 0, and we suppose that it changes in accordance with a Wiener process having a negative drift coefficient $-d$, where $d > 0$. The question is when, if ever, should we exercise our option?

Let us consider the policy which exercises the option when the market price is x. Our expected gain under such a policy is

$$P(x)(x - A) \qquad (6)$$

where $P(x)$ is the probability that a Wiener process with drift $-d$ will ever reach x. Hence, we must calculate $P(x)$. To do so, we first note that in order to reach $x + y$, where x and y are positive, we must first reach x and then from x we must reach $x + y$. However, from the assumption of stationary independent increments, it follows that the probability that the Wiener process ever reaches a height of $x + y$, given that it has reached a height of x, is just $P(y)$. Hence,

$$P(x + y) = P(x)P(y)$$

and thus,

$$P(x) = e^{-\lambda x}, \qquad x > 0 \qquad (7)$$

It now remains to calculate λ. In order to do this, we condition on the value of the process at time h, where h is small, Arguing heuristically, we have

$$P(x) = E[P(x - Y)] + o(h)$$

where Y is a normal random variable with mean $-dh$ and variance h. That is, Y represents the market price at time h. Proceeding formally and applying Taylor's formula yields

$$P(x) = E[P(x) - P'(x)Y + P''(x)Y^2/2 + \cdots] + o(h)$$
$$= P(x) + dhP'(x) + P''(x)(h + d^2h^2)/2 + \cdots + o(h)$$

Now, by using (7), we obtain

$$0 = -\lambda h d e^{-\lambda x} + \lambda^2 e^{-\lambda x}(h + d^2h^2)/2 + o(h)$$

Now, dividing by h, and then letting $h \to 0$ yields

$$d = \lambda/2$$

Hence,

$$P(x) = e^{-2dx}, \qquad x > 0$$

The optimal value of x is the one maximizing $(x - A)e^{-2dx}$, and this is easily seen to be

$$x = A + \frac{1}{2d}$$

EXAMPLE 2. *Controlling a Production Process.* In this example, we consider a production process which tends to deteriorate with time. Specifically, we suppose that the production process changes its state in accordance with a Wiener process with drift coefficient μ, $\mu > 0$. When the state of the process is B, the process is assumed to break down and a cost R must be paid to return the process back to state 0. On the other hand, we may attempt to repair the process before it reaches the breakdown point B. If the state is x and an attempt to repair the process is made, then this attempt will succeed with probability α_x and fail with probability $1 - \alpha_x$. If the attempt is successful, then the process returns to state 0, and if it is unsuccessful, then we assume that the process goes to B (that is, it breaks down). The cost of attempting a repair is C.

We shall attempt to determine the policy with minimizes the long-run average cost per time, and in doing so we will restrict attention to policies which attempt a repair when the state of the process is x, $0 < x < B$. For these policies, it is clear that returns to state 0 constitute renewals, and thus by Theorem 3.16, the average cost is just

$$\frac{E[\text{Cost of a cycle}]}{E[\text{Length of a cycle}]} = \frac{C + R(1 - \alpha_x)}{E[\text{Time to reach } x]} \tag{8}$$

Now, let $f(x)$ denote the expected time that it takes the process to reach x. We first note that due to the stationary independent increment assumption, we have

$$f(x + y) = f(x) + f(x)$$

and hence,

$$f(x) = cx$$

In order to calculate the constant c, we let

$$X_1 = X(h)$$
$$X_2 = X(2h) - X(h)$$
$$\vdots$$
$$X_n = X(nh) - X[(n - 1)h]$$

and let N be the smallest n such that $\sum_{i=1}^{n} X_i \geq x$. Then, by Wald's equation (Theorem 3.6),

$$E \sum_{i=1}^{N} X_i = EN \cdot h\mu$$

but $\sum_{i=1}^{N} X_i \approx x$, and this approximation becomes exact as $h \to 0$. Hence,

$$EN \approx \frac{x}{h\mu}$$

However, if we let T denote the time at which the process first reaches x, then $T \approx Nh$, and hence,

$$ET \approx \frac{hx}{h\mu}$$

and by letting $h \to 0$, we obtain,

$$f(x) = ET = \frac{x}{\mu} \qquad (9)$$

Hence, from (8), the policy which attempts to repair when the state is x, $0 < x < B$, has a long-run average cost of

$$\frac{\mu[C + R(1 - \alpha_x)]}{x}$$

while the policy which never attempts to repair has a long-run average cost of

$$\frac{R\mu}{B}$$

Thus, for instance, if $1 - \alpha_x = x/B$, then the optimal policy would never attempt a repair; while if

$$\alpha_x = \begin{cases} \dfrac{e^{-\lambda x} - e^{-\lambda B}}{1 - e^{-\lambda B}} & 0 < x < B \\ 0 & x \geq B \end{cases}$$

then (assuming that a repair is optimal) the optimal policy attempts a repair at that value of x satisfying

$$\lambda x e^{-\lambda x} + e^{-\lambda x} - 1 = \frac{C(1 - e^{-\lambda B})}{R}$$

9.4. The Maximum Variable—A Renewal Application

Consider a Wiener process $\{X(t), t \geq 0\}$ having a drift coefficient μ, where $\mu \geq 0$. In this section we shall investigate the limiting behavior of

$$\frac{\max_{0 \leq s \leq t} X(s)}{t}$$

Specifically, we prove

Theorem 9.1

If $\{X(t), t \geq 0\}$ is a Wiener process with drift $\mu \geq 0$, then, with probability 1,

$$\lim_{t \to \infty} \frac{\max_{0 \leq s \leq t} X(s)}{t} = \mu$$

PROOF. Let $T_0 = 0$, and for $n > 0$ let T_n denote the time at which the process hits n. It follows, from the assumption of stationary, independent increments, that $T_n - T_{n-1}, n \geq 1$, are independent and identically distributed. Hence, we may think of the T_n as being the times at which events occur in a renewal process. Letting $N(t)$ be the number of such renewals by t, we have that

$$N(t) \leq \max_{0 \leq s \leq t} X(s) \leq N(t) + 1 \tag{10}$$

Now, from (9), we have that $ET_1 = 1/\mu$, and hence the result follows from (10) and the well-known renewal result that $N(t)/t \to 1/ET_1$ (Theorem 3.5).

9.5. Optimal Dispatching of a Poisson Process

Suppose that items arrive at a processing plant in accordance with a Poisson process with rate λ. At a fixed time T, all items are dispatched from the system. The problem is to choose an intermediate dispatch time, when all items in the system are dispatched, so as to minimize the expected total wait of all items. In Section 2.3 this problem was considered under the restriction that the intermediate dispatch time be constant. However, in this section we allow it to be a random stopping time. That is, we allow the decision on whether or not to stop at time t to depend on $\{N(s), s \leq t\}$, where $N(s)$ denotes the total number of arrivals by time s.

We say that a policy (i.e., stopping time) is optimal if it minimizes the expected total wait. Let δ be the policy which dispatches at

$$\tau = \min\{t \geq 0 : N(t) \geq \lambda(T - t)\}.$$

That is, δ dispatches at the first time that the number of items present is greater than or equal to the expected number to arrive from then on.

Note that $N(\tau) = \lambda(T - \tau)$ unless $N(t)$ has a jump at τ. (See Figure 9.1)

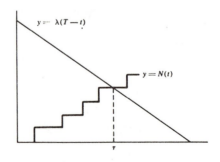

Figure 9.1.

Theorem 9.2

δ *is optimal.*

PROOF. Let δ_1 be any policy, and let δ_2 follow δ_1, with the exception that if $N(t) < \lambda(T - t)$ and δ_1 dispatches at t, then δ_2 dispatches at

$$t + (\lambda(T - t) - N(t))/\lambda \equiv t + h.$$

The expected total wait from time t onward under δ_1 is $\lambda(T - t)^2/2$, while the expected total wait from t onward under δ_2 is

$$N(t) \cdot h + \frac{\lambda h^2}{2} + \frac{\lambda(T - t - h)^2}{2} = \frac{\lambda(T - t)^2}{2}. \tag{11}$$

and thus δ_1 and δ_2 have the same expected wait. [The reasoning behind (11) is that the $N(t)$ items in the system at t will have to wait an additional h units of time, the expected wait for items arriving in $(t, t + h)$ is $\lambda h^2/2$ and the expected wait for items arriving in $(t + h, T)$ is $\lambda(T - t - h)^2/2$.]

Now let δ denote the policy which dispatches at

$$\tau = \min \{t \geq 0 : N(t) \geq \lambda(T - t)\}$$

Since $N(t + h) \geq N(t) = \lambda[T - (t + h)]$, it follows from the definition of δ_2 that $t(\delta_2) \geq \tau$, where $t(\delta_2)$ denotes the time at which δ_2 dispatches. Thus, $t(\delta_2) = \tau + \varepsilon$, where ε is nonnegative and random. Hence, the conditional expected total wait from τ under δ_2 is at least

$$N(\tau) \cdot \varepsilon + \frac{\lambda(T - \tau - \varepsilon)^2}{2} \geq \lambda(T - \tau)\varepsilon + \frac{\lambda(T - \tau - \varepsilon)^2}{2} = \frac{\lambda(T - \tau)^2}{2} + \frac{\lambda\varepsilon^2}{2}$$

which is strictly greater than $\lambda(T - \tau)^2/2 (= $ the conditional expected total wait from τ under δ) whenever $\varepsilon > 0$. From this, it follows that δ is at least as good as δ_2, and hence at least as good as δ_1, and the theorem is proven.

9.6. Infinitesimal Look-Ahead Stopping Rules

Let us now view the optimal dispatch problem in a slightly different manner. Suppose that there are $N(t)$ people in the system at t. Then, if we dispatch at t, the total wait from t is

$$\frac{\lambda(T - t)^2}{2}$$

while if we dispatch at $t + \varepsilon$ (ε fixed and small), then the total wait from t is

$$N(t) \cdot \varepsilon + \frac{\lambda\varepsilon^2}{2} + \frac{\lambda(T - t - \varepsilon)^2}{2}$$

Hence, dispatching at t is better than waiting an additional small time ε if

$$\frac{\lambda(T-t)^2}{2} < N(t) \cdot \varepsilon + \frac{\lambda\varepsilon^2}{2} + \frac{\lambda(T-t-\varepsilon)^2}{2}$$

or equivalently, if

$$0 < N(t) \cdot \varepsilon + \lambda\varepsilon^2 - \lambda\varepsilon(T-t)$$

or

$$\lambda(T-t) < N(t) + \lambda\varepsilon$$

Thus, dispatching at t is better than waiting an additional small time ε if and only if

$$N(t) \geq \lambda(T-t)$$

Now, at any time t, the state of the process may be completely specified by the pair $[t, N(t)]$. Hence, if we define the *infinitesimal look-ahead* (*ILA*) stopping rule to be the policy which stops (i.e., dispatches) at those states for which stopping (i.e., dispatching) immediately is better than waiting an additional small time ε, then for the optimal dispatch problem just considered the *ILA* rule is optimal.

Clearly, the *ILA* rule is the continuous time analog of the one stage look ahead rule in the discrete time stopping problem (see Section 6.5). Hence, if we define B as the set of states for which stopping immediately is better than waiting an additional small time ε, then by analogy with Theorem 6.14 it seems that we should be able to prove a theorem such as

Theorem 9.3

If B is closed in the sense that once the process enters B, then it cannot leave it, then, under certain regularity conditions, the ILA rule is optimal.

It turns out that such a theorem is indeed true (see [4] for a proof), and we shall accept it without proof (and without specifying the necessary regularity conditions).†

To show how Theorem 9.3 works, first consider our dispatch problem. As previously shown,

$$B = \{[N(t), t] : N(t) \geq \lambda(T-t)\}$$

† Basically, the regularity conditions are necessary to ensure the continuous time version of stability (see Section 6.5).

Now, since $N(t)$ never decreases and $\lambda(T - t)$ always decreases, it follows that once the process enters B it remains there. Hence, by Theorem 9.3, we have the previously proven result that it is optimal to dispatch at t if and only if $N(t) \geq \lambda(T - t)$.

We shall now consider some other examples illustrating the usefulness of Theorem 9.3.

EXAMPLE 3.   *A House-Selling Example.*   Suppose that offers to buy a house are made in accordance with a nonhomogeneous Poisson process with a continuous nonincreasing intensity function $\lambda(t)$. The offers are independent and have a common distribution F with finite mean. Suppose also that any offer not immediately accepted is not lost but may be accepted at any later date. Maintenance costs at a rate c per unit time are incurred until the house is sold.

The state of the process at any time is given by the pair (t, m), where t represents the time and m the maximum value offered by t. If we sell when in (t, m), then our return (from that point on) is m, while if we wait an additional time ε, then our expected return is

$$\left(1 - \int_t^{t+\varepsilon} \lambda(s)\, ds\right)m + \int_t^{t+\varepsilon} \lambda(s)\, ds \cdot E[\max(X, m)] - \varepsilon c + o(\varepsilon)$$

$$= m + \int_t^{t+\varepsilon} \lambda(s)\, ds \int_m^\infty (x - m)\, dF(x) - \varepsilon c + o(\varepsilon)$$

Hence, stopping immediately is better if (for ε small)

$$0 \geq \int_t^{t+\varepsilon} \lambda(s)\, ds \int_m^\infty (x - m)\, dF(x) - \varepsilon c + o(\varepsilon)$$

or equivalently, if

$$c \geq \lambda(t) \int_m^\infty (x - m)\, dF(x)$$

Hence,

$$B = \left\{(t, m) : c \geq \lambda(t) \int_m^\infty (x - m)\, dF(x)\right\}$$

Since as t increases, $\lambda(t)$ decreases and $\int_m^\infty (x - m)\, dF(x)$ does not increase (since m can't get smaller), it follows that the conditions of Theorem 9.3 hold; hence, the optimal policy is to sell at t if and only if the maximal offer by t is at least $m(t)$, where $m(t)$ is such that

$$c = \lambda(t) \int_{m(t)}^\infty [x - m(t)]\, dF(x)$$

EXAMPLE 4.   Let $\{N(t), t \geq 0\}$ be a Poisson process with rate λ. Suppose that the reward for stopping when $N(t) = n$ is n, and the continuation rate is c. Suppose further that we continuously discount returns (and costs). That is, suppose that a value of A at time t is worth $Ae^{-\alpha t}$ at time 0.

If we stop when in state n, then our return is n, while if we continue for a time ε, then our expected return is

$$e^{-\alpha \varepsilon}[n(1 - \lambda \varepsilon) + (n + 1)\lambda \varepsilon + o(\varepsilon)] - c \int_0^\varepsilon e^{-\lambda t} \, dt = n - n\alpha\varepsilon + \lambda\varepsilon + o(\varepsilon) - c\varepsilon$$

Hence,

$$B = \left\{ n : n \geq \frac{\lambda - c}{\alpha} \right\}$$

and since $N(t)$ cannot decrease, it follows from Theorem 9.3 that the optimal policy stops at time t if and only if $N(t) \geq (\lambda - c)/\alpha$.

Problems

1.   Prove Equations (2) and (3).

2.   Let $\{X(t), t \geq 0\}$ be a Wiener process with $\mu = 0$. Show that the following pairs of random variables have the same distribution:

  (i) $X(t)$ and $tX(1/t)$;
  (ii) $X(t)$ and $1/aX(a^2t)$, $a > 0$.

3.   For a Wiener process with $\mu = 0$, find the probability that $X(t)$ has at least one zero between s and $s + a$, $a > 0$.
Hint: Condition on $X(s)$.

4.   Calculate the probability that a Wiener process with drift coefficient μ reaches $-a$ before it reaches b.

5.   Let $X(t)$ have drift μ. Prove that

$$\lim_{t \to \infty} \frac{X(t)}{t} = \mu \qquad \text{with probability 1}$$

6.   Consider an individual who owns a stock whose present value is A. Suppose that changes in the market value occur in accordance with a Wiener process with nonnegative drift coefficient μ. The individual has until time T to sell his stock. What is his optimal policy? (The stock must be sold at T if he has not sold it by then.)

7.   Consider the above problem under the assumption that a return of x at time t is only worth $xe^{-\alpha t}$ at time 0.

8.   Let $\{N(t), t \geq 0\}$ be a Poisson process with rate λ, and let τ be a stopping time. That is, the event that $\tau < s$ is independent of $\{N(t) - N(s), t > s\}$. Show that $E[N(\tau)] = \lambda E\tau$.

9.   Find the distribution for the stopping time τ of Section 9.5.

10. Find the expected wait under the optimal policy of Section 9.5.

11. Solve the optimal dispatch problem when the arrival stream is a nonhomogeneous Poisson process with a continuous, nonincreasing rate function $\lambda(t)$.

12. Solve the model of Section 9.5 under the assumption that each arrival consists of a random number of customers.

13. Solve the optimal dispatch problem when there is a time lag L between the time at which the decision to dispatch is made and the time of actual dispatch.

References

Example 1 was taken from Taylor [5] where other stock option models are also considered. Example 2, the production process example, is similar to models considered by Antelman and Savage [1]. The optimal dispatching example of Section 9.5 is due to Ross [3]. The results of Section 9.6 are also due to Ross [4].

For further readings on the Wiener process, the interested reader should consult Lévy [2].

[1] ANTELMAN, G. and I. R. SAVAGE, "Surveillance Problems: Wiener Process," *Naval Research Logistics Quarterly*, **12**, pp. 35–55, (1965).

[2] LÉVY, P. *Processus Stochastiques et Mouvement Brownian.* Gauthier-Villars, Paris, (1948).

[3] ROSS, S. "Optimal Dispatching of a Poisson Process," to appear in *Journal of Applied Probability*, (1969).

[4] ROSS, S. "Infinitesimal Look-Ahead Stopping Rules," Operations Research Center Report ORC 69-7, University of California, Berkeley, (1969).

[5] TAYLOR, H. "Evaluating a Call Option and Optimal Timing Strategy in the Stock Market," *Management Science*, **12**, pp. 111–120, (1967–68).

APPENDICES

Appendix 1. Contraction Mappings

Let X be a metric space with metric ρ. That is, $\rho(x, y)$ represents the *distance* from x to y, and it satisfies:

  (i) $\rho(x, y) = \rho(y, x)$;
  (ii) $\rho(x, y) = 0$ if and only if $x = y$;
  (iii) $\rho(x, y) \leq \rho(x, z) + \rho(z, y)$.

The space X is said to be *complete* if for every sequence $\{x_n\}$ such that $\rho(x_n, x_{n+m}) \to 0$ as $n \to \infty$, for each m, there exists an $x \in X$ such that $\rho(x_n, x) \to 0$. We say that $x = \lim x_n$.

An operator T mapping X into itself is called a *contraction mapping* if for some $\beta \in (0, 1)$,

$$\rho(Tx, Ty) \leq \beta\rho(x, y) \qquad \text{for all } x, y \in X.$$

Theorem: (*Fixed Point Theorem for Contraction Mappings*)

  if X is a complete metric space, and T a contraction mapping, then there exists a unique point v such that

$$Tv = v$$

Furthermore, for any $x \in X$,

$$\rho(T^n x, v) \to 0$$

where $T^1 x = Tx$, $T^n x = T(T^{n-1} x)$.

PROOF. We first prove uniqueness. This follows, since if $Tu = u$, $Tv = v$, then

$$\rho(u, v) = \rho(Tu, Tv) \leq \beta\rho(u, v)$$

which implies that $\rho(u, v) = 0$ and hence $u = v$. Now, for any x, consider $\rho(T^{n+m}x, T^n x)$. We have

$$
\begin{aligned}
\rho(T^{n+m}x, T^n x) &\le \beta^n \rho(T^m x, x) \\
&\le \beta^n [\rho(T^m x, T^{m-1}x) + \cdots + \rho(Tx, x)] \\
&\le \beta^n \rho(Tx, x)(1 + \beta + \cdots + \beta^{m+1}) \\
&\to 0 \quad \text{as} \quad n \to \infty
\end{aligned}
$$

Hence from completeness we obtain that $v \equiv \lim T^n x$ exists. Now T is a continuous mapping, since if $\rho(x_n, x) \to 0$, then $\rho(Tx_n, Tx) \le \beta \rho(x_n, x) \to 0$. As a result,

$$
Tv = T \lim_n T^n x = \lim_n T^{n+1}x = v
$$

Q.E.D.

Appendix 2. A Counterexample

The following is an example for which there is an optimal nonstationary policy which is better than every stationary policy.

EXAMPLE. The state space will be denoted by the set $\{0, 1, 1', 2, 2', \dots\}$. There are two actions and the transition probabilities are given by

$$
\begin{aligned}
P_{0i}(j) = P_{0i'}(j) &= \tfrac{3}{2}(\tfrac{1}{4})^i, \qquad i > 0, j = 1, 2 \\
P_{i0}(1) = (\tfrac{1}{2})^i &= 1 - P_{ii'}(1), \\
P_{i0}(2) = \tfrac{1}{2} &= 1 - P_{ii+1}(2), \\
P_{i'0}(j) = (\tfrac{1}{2})^i &= 1 - P_{i'i'}(j), \qquad j = 1, 2
\end{aligned}
$$

The costs are zero except when in state 0. That is,

$$
\begin{aligned}
C(0, \cdot) &= 1 \\
C(i, \cdot) = C(i', \cdot) &= 0, \qquad i > 0
\end{aligned}
$$

Let π_n be the stationary deterministic rule which takes action 2 at states $0 < i < n$ and action 1 elsewhere, and let $M_{j0}(\pi_n)$ be the mean recurrence time when π_n is used. Then,

$$
M_{00}(\pi_n) = 1 + \sum_{j=1}^{\infty} \tfrac{3}{2}(\tfrac{1}{4})^j M_{j0}(\pi_n) + \sum_{j=1}^{\infty} \tfrac{3}{2}(\tfrac{1}{4})^j 2^j
$$

Now $j \ge n \Rightarrow M_{j0}(\pi_n) = 2^j$, whereas

$$
\begin{aligned}
j < n \Rightarrow M_{j0}(\pi_n) &= \tfrac{1}{2} + 2(\tfrac{1}{2})^2 + \cdots + (n-j)(\tfrac{1}{2})^{n-j} + (\tfrac{1}{2})^{n-j}[n - j + 2^n] \\
&= 2 + 2^j - (\tfrac{1}{2})^{n-j-1}
\end{aligned}
$$

Therefore,

$$M_{00}(\pi_n) = \tfrac{5}{2} + \sum_{j=1}^{\infty} \tfrac{3}{2}(\tfrac{1}{4})^j(2 + 2^j) - \sum_{j=1}^{n-1} \tfrac{3}{2}(\tfrac{1}{4})^j(\tfrac{1}{2})^{n-j-1}$$
$$- 2\sum_{j=n}^{\infty} \tfrac{3}{2}(\tfrac{1}{4})^j$$
$$= 5 - \sum_{j=1}^{n-1} \tfrac{3}{2}(\tfrac{1}{4})^j(\tfrac{1}{2})^{n-j-1} - 3\sum_{j=n}^{\infty} (\tfrac{1}{4})^j$$

Hence $M_{00}(\pi_n) < 5$ for all n, and $M_{00}(\pi_n) \to 5$ as $n \to \infty$.

Now let π be any stationary rule, let P_i be the probability that π takes action 1 when in state i. Now,

$$M_{00}(\pi) = 1 + \sum_{j=1}^{\infty} \tfrac{3}{2}(\tfrac{1}{4})^j M_{j0}(\pi) + \sum_{j=1}^{\infty} \tfrac{3}{2}(\tfrac{1}{4})^j 2^j$$

but

$$M_{j0}(\pi) = \sum_{n=j}^{\infty} \left[P_n \prod_{k=j}^{n-1} (1 - P_k) \right] M_{j0}(\pi_n) + 2 \prod_{k=j}^{\infty} (1 - P_k)$$
$$< (2 + 2^j) \left[\sum_{n=j}^{\infty} P_n \prod_{k=j}^{n-1}(1 - P_k) + \prod_{k=j}^{\infty}(1 - P_k) \right] = 2 + 2^j$$

Consequently,

$$M_{00}(\pi) < 5 \qquad \text{for all stationary rules } \pi$$

However, since the average cost is just the ratio of the expected cost incurred during a cycle to the expected length of the cycle, this implies that

$$\phi_\pi(i) > \tfrac{1}{5} \qquad \text{for all stationary policies, all } i$$

However, if we consider the nonstationary policy π^* which uses

$$\pi_1 \quad \text{for} \quad t = 1, 2, \ldots, N_1$$
$$\pi_2 \quad \text{for} \quad t = N_1 + 1, \ldots, N_1 + N_2$$
$$\vdots$$
$$\pi_n \quad \text{for} \quad t = \sum_{i=1}^{n-1} N_i + 1, \ldots, \sum_{i=1}^{n} N_i$$

then it can be shown (see [1]) that there exist N_j's such that

$$\phi_{\pi^*}(i) = \lim \phi_{\pi_n}(i) = \tfrac{1}{5}.$$

Reference

FISHER, L. and S. ROSS, "An Example in Denumerable Decision Processes," *Annals of Mathematical Statistics*, **39**, pp. 674–675, (1968).

INDEX

A CATALOG OF SELECTED
DOVER BOOKS
IN SCIENCE AND MATHEMATICS

A CATALOG OF SELECTED

DOVER BOOKS

IN SCIENCE AND MATHEMATICS

QUALITATIVE THEORY OF DIFFERENTIAL EQUATIONS, V.V. Nemytskii and V.V. Stepanov. Classic graduate-level text by two prominent Soviet mathematicians covers classical differential equations as well as topological dynamics and ergodic theory. Bibliographies. 523pp. 5⅜ x 8½. 65954-2 Pa. $14.95

MATRICES AND LINEAR ALGEBRA, Hans Schneider and George Phillip Barker. Basic textbook covers theory of matrices and its applications to systems of linear equations and related topics such as determinants, eigenvalues and differential equations. Numerous exercises. 432pp. 5⅜ x 8½. 66014-1 Pa. $10.95

QUANTUM THEORY, David Bohm. This advanced undergraduate-level text presents the quantum theory in terms of qualitative and imaginative concepts, followed by specific applications worked out in mathematical detail. Preface. Index. 655pp. 5⅜ x 8½. 65969-0 Pa. $14.95

ATOMIC PHYSICS (8th edition), Max Born. Nobel laureate's lucid treatment of kinetic theory of gases, elementary particles, nuclear atom, wave-corpuscles, atomic structure and spectral lines, much more. Over 40 appendices, bibliography. 495pp. 5⅜ x 8½. 65984-4 Pa. $13.95

ELECTRONIC STRUCTURE AND THE PROPERTIES OF SOLIDS: The Physics of the Chemical Bond, Walter A. Harrison. Innovative text offers basic understanding of the electronic structure of covalent and ionic solids, simple metals, transition metals and their compounds. Problems. 1980 edition. 582pp. 6⅛ x 9¼. 66021-4 Pa. $16.95

BOUNDARY VALUE PROBLEMS OF HEAT CONDUCTION, M. Necati Özisik. Systematic, comprehensive treatment of modern mathematical methods of solving problems in heat conduction and diffusion. Numerous examples and problems. Selected references. Appendices. 505pp. 5⅜ x 8½. 65990-9 Pa. $12.95

A SHORT HISTORY OF CHEMISTRY (3rd edition), J.R. Partington. Classic exposition explores origins of chemistry, alchemy, early medical chemistry, nature of atmosphere, theory of valency, laws and structure of atomic theory, much more. 428pp. 5⅜ x 8½. (Available in U.S. only) 65977-1 Pa. $11.95

A HISTORY OF ASTRONOMY, A. Pannekoek. Well-balanced, carefully reasoned study covers such topics as Ptolemaic theory, work of Copernicus, Kepler, Newton, Eddington's work on stars, much more. Illustrated. References. 521pp. 5⅜ x 8½. 65994-1 Pa. $12.95

PRINCIPLES OF METEOROLOGICAL ANALYSIS, Walter J. Saucier. Highly respected, abundantly illustrated classic reviews atmospheric variables, hydrostatics, static stability, various analyses (scalar, cross-section, isobaric, isentropic, more). For intermediate meteorology students. 454pp. 6⅛ x 9¼. 65979-8 Pa. $14.95

CATALOG OF DOVER BOOKS

RELATIVITY, THERMODYNAMICS AND COSMOLOGY, Richard C. Tolman. Landmark study extends thermodynamics to special, general relativity; also applications of relativistic mechanics, thermodynamics to cosmological models. 501pp. 5⅜ x 8½. 65383-8 Pa. $13.95

APPLIED ANALYSIS, Cornelius Lanczos. Classic work on analysis and design of finite processes for approximating solution of analytical problems. Algebraic equations, matrices, harmonic analysis, quadrature methods, much more. 559pp. 5⅜ x 8½. 65656-X Pa. $13.95

INTRODUCTION TO ANALYSIS, Maxwell Rosenlicht. Unusually clear, accessible coverage of set theory, real number system, metric spaces, continuous functions, Riemann integration, multiple integrals, more. Wide range of problems. Undergraduate level. Bibliography. 254pp. 5⅜ x 8½. 65038-3 Pa. $8.95

INTRODUCTION TO QUANTUM MECHANICS With Applications to Chemistry, Linus Pauling & E. Bright Wilson, Jr. Classic undergraduate text by Nobel Prize winner applies quantum mechanics to chemical and physical problems. Numerous tables and figures enhance the text. Chapter bibliographies. Appendices. Index. 468pp. 5⅜ x 8½. 64871-0 Pa. $12.95

ASYMPTOTIC EXPANSIONS OF INTEGRALS, Norman Bleistein & Richard A. Handelsman. Best introduction to important field with applications in a variety of scientific disciplines. New preface. Problems. Diagrams. Tables. Bibliography. Index. 448pp. 5⅜ x 8½. 65082-0 Pa. $12.95

MATHEMATICS APPLIED TO CONTINUUM MECHANICS, Lee A. Segel. Analyzes models of fluid flow and solid deformation. For upper-level math, science and engineering students. 608pp. 5⅜ x 8½. 65369-2 Pa. $14.95

ELEMENTS OF REAL ANALYSIS, David A. Sprecher. Classic text covers fundamental concepts, real number system, point sets, functions of a real variable, Fourier series, much more. Over 500 exercises. 352pp. 5⅜ x 8½. 65385-4 Pa. $11.95

PHYSICAL PRINCIPLES OF THE QUANTUM THEORY, Werner Heisenberg. Nobel Laureate discusses quantum theory, uncertainty, wave mechanics, work of Dirac, Schroedinger, Compton, Wilson, Einstein, etc. 184pp. 5⅜ x 8½. 60113-7 Pa. $6.95

INTRODUCTORY REAL ANALYSIS, A.N. Kolmogorov, S.V. Fomin. Translated by Richard A. Silverman. Self-contained, evenly paced introduction to real and functional analysis. Some 350 problems. 403pp. 5⅜ x 8½. 61226-0 Pa. $10.95

PROBLEMS AND SOLUTIONS IN QUANTUM CHEMISTRY AND PHYSICS, Charles S. Johnson, Jr. and Lee G. Pedersen. Unusually varied problems, detailed solutions in coverage of quantum mechanics, wave mechanics, angular momentum, molecular spectroscopy, scattering theory, more. 280 problems plus 139 supplementary exercises. 430pp. 6½ x 9¼. 65236-X Pa. $13.95

ASYMPTOTIC METHODS IN ANALYSIS, N.G. de Bruijn. An inexpensive, comprehensive guide to asymptotic methods–the pioneering work that teaches by explaining worked examples in detail. Index. 224pp. 5⅜ x 8½.   64221-6 Pa. $7.95

OPTICAL RESONANCE AND TWO-LEVEL ATOMS, L. Allen and J. H. Eberly. Clear, comprehensive introduction to basic principles behind all quantum optical resonance phenomena. 53 illustrations. Preface. Index. 256pp. 5⅜ x 8½.
65533-4 Pa. $8.95

COMPLEX VARIABLES, Francis J. Flanigan. Unusual approach, delaying complex algebra till harmonic functions have been analyzed from real variable viewpoint. Includes problems with answers. 364pp. 5⅜ x 8½.   61388-7 Pa. $9.95

ATOMIC SPECTRA AND ATOMIC STRUCTURE, Gerhard Herzberg. One of best introductions; especially for specialist in other fields. Treatment is physical rather than mathematical. 80 illustrations. 257pp. 5⅜ x 8½.   60115-3 Pa. $7.95

APPLIED COMPLEX VARIABLES, John W. Dettman. Step-by-step coverage of fundamentals of analytic function theory–plus lucid exposition of five important applications: Potential Theory; Ordinary Differential Equations; Fourier Transforms; Laplace Transforms; Asymptotic Expansions. 66 figures. Exercises at chapter ends. 512pp. 5⅜ x 8½.   64670-X Pa. $12.95

ULTRASONIC ABSORPTION: An Introduction to the Theory of Sound Absorption and Dispersion in Gases, Liquids and Solids, A.B. Bhatia. Standard reference in the field provides a clear, systematically organized introductory review of fundamental concepts for advanced graduate students, research workers. Numerous diagrams. Bibliography. 440pp. 5⅜ x 8½.   64917-2 Pa. $11.95

UNBOUNDED LINEAR OPERATORS: Theory and Applications, Seymour Goldberg. Classic presents systematic treatment of the theory of unbounded linear operators in normed linear spaces with applications to differential equations. Bibliography. 199pp. 5⅜ x 8½.   64830-3 Pa. $7.95

LIGHT SCATTERING BY SMALL PARTICLES, H.C. van de Hulst. Comprehensive treatment including full range of useful approximation methods for researchers in chemistry, meteorology and astronomy. 44 illustrations. 470pp. 5⅜ x 8½.
64228-3 Pa. $12.95

CONFORMAL MAPPING ON RIEMANN SURFACES, Harvey Cohn. Lucid, insightful book presents ideal coverage of subject. 334 exercises make book perfect for self-study. 55 figures. 352pp. 5⅜ x 8¼.   64025-6 Pa. $11.95

OPTICKS, Sir Isaac Newton. Newton's own experiments with spectroscopy, colors, lenses, reflection, refraction, etc., in language the layman can follow. Foreword by Albert Einstein. 532pp. 5⅜ x 8½.   60205-2 Pa. $12.95

GENERALIZED INTEGRAL TRANSFORMATIONS, A.H. Zemanian. Graduate-level study of recent generalizations of the Laplace, Mellin, Hankel, K. Weierstrass, convolution and other simple transformations. Bibliography. 320pp. 5⅜ x 8½.
65375-7 Pa. $8.95

THE ELECTROMAGNETIC FIELD, Albert Shadowitz. Comprehensive undergraduate text covers basics of electric and magnetic fields, builds up to electromagnetic theory. Also related topics, including relativity. Over 900 problems. 768pp. 5⅜ x 8¼. 65660-8 Pa. $18.95

FOURIER SERIES, Georgi P. Tolstov. Translated by Richard A. Silverman. A valuable addition to the literature on the subject, moving clearly from subject to subject and theorem to theorem. 107 problems, answers. 336pp. 5⅜ x 8½. 63317-9 Pa. $9.95

THEORY OF ELECTROMAGNETIC WAVE PROPAGATION, Charles Herach Papas. Graduate-level study discusses the Maxwell field equations, radiation from wire antennas, the Doppler effect and more. xiii + 244pp. 5⅜ x 8½. 65678-0 Pa. $6.95

DISTRIBUTION THEORY AND TRANSFORM ANALYSIS: An Introduction to Generalized Functions, with Applications, A.H. Zemanian. Provides basics of distribution theory, describes generalized Fourier and Laplace transformations. Numerous problems. 384pp. 5⅜ x 8½. 65479-6 Pa. $11.95

THE PHYSICS OF WAVES, William C. Elmore and Mark A. Heald. Unique overview of classical wave theory. Acoustics, optics, electromagnetic radiation, more. Ideal as classroom text or for self-study. Problems. 477pp. 5⅜ x 8½. 64926-1 Pa. $13.95

CALCULUS OF VARIATIONS WITH APPLICATIONS, George M. Ewing. Applications-oriented introduction to variational theory develops insight and promotes understanding of specialized books, research papers. Suitable for advanced undergraduate/graduate students as primary, supplementary text. 352pp. 5⅜ x 8½. 64856-7 Pa. $9.95

A TREATISE ON ELECTRICITY AND MAGNETISM, James Clerk Maxwell. Important foundation work of modern physics. Brings to final form Maxwell's theory of electromagnetism and rigorously derives his general equations of field theory. 1,084pp. 5⅜ x 8½. 60636-8, 60637-6 Pa., Two-vol. set $25.90

AN INTRODUCTION TO THE CALCULUS OF VARIATIONS, Charles Fox. Graduate-level text covers variations of an integral, isoperimetrical problems, least action, special relativity, approximations, more. References. 279pp. 5⅜ x 8½. 65499-0 Pa. $8.95

HYDRODYNAMIC AND HYDROMAGNETIC STABILITY, S. Chandrasekhar. Lucid examination of the Rayleigh-Benard problem; clear coverage of the theory of instabilities causing convection. 704pp. 5⅜ x 8¼. 64071-X Pa. $14.95

CALCULUS OF VARIATIONS, Robert Weinstock. Basic introduction covering isoperimetric problems, theory of elasticity, quantum mechanics, electrostatics, etc. Exercises throughout. 326pp. 5⅜ x 8½. 63069-2 Pa. $9.95

DYNAMICS OF FLUIDS IN POROUS MEDIA, Jacob Bear. For advanced students of ground water hydrology, soil mechanics and physics, drainage and irrigation engineering and more. 335 illustrations. Exercises, with answers. 784pp. 6⅛ x 9¼. 65675-6 Pa. $19.95

NUMERICAL METHODS FOR SCIENTISTS AND ENGINEERS, Richard Hamming. Classic text stresses frequency approach in coverage of algorithms, polynomial approximation, Fourier approximation, exponential approximation, other topics. Revised and enlarged 2nd edition. 721pp. 5⅜ x 8½. 65241-6 Pa. $15.95

THEORETICAL SOLID STATE PHYSICS, Vol. 1: Perfect Lattices in Equilibrium; Vol. II: Non-Equilibrium and Disorder, William Jones and Norman H. March. Monumental reference work covers fundamental theory of equilibrium properties of perfect crystalline solids, non-equilibrium properties, defects and disordered systems. Appendices. Problems. Preface. Diagrams. Index. Bibliography. Total of 1,301pp. 5⅜ x 8½. Two volumes. Vol. I: 65015-4 Pa. $16.95
Vol. II: 65016-2 Pa. $16.95

OPTIMIZATION THEORY WITH APPLICATIONS, Donald A. Pierre. Broad spectrum approach to important topic. Classical theory of minima and maxima, calculus of variations, simplex technique and linear programming, more. Many problems, examples. 640pp. 5⅜ x 8½. 65205-X Pa. $16.95

THE CONTINUUM: A Critical Examination of the Foundation of Analysis, Hermann Weyl. Classic of 20th-century foundational research deals with the conceptual problem posed by the continuum. 156pp. 5⅜ x 8½. 67982-9 Pa. $6.95

ESSAYS ON THE THEORY OF NUMBERS, Richard Dedekind. Two classic essays by great German mathematician: on the theory of irrational numbers; and on transfinite numbers and properties of natural numbers. 115pp. 5⅜ x 8½. 21010-3 Pa. $5.95

THE FUNCTIONS OF MATHEMATICAL PHYSICS, Harry Hochstadt. Comprehensive treatment of orthogonal polynomials, hypergeometric functions, Hill's equation, much more. Bibliography. Index. 322pp. 5⅜ x 8½. 65214-9 Pa. $9.95

NUMBER THEORY AND ITS HISTORY, Oystein Ore. Unusually clear, accessible introduction covers counting, properties of numbers, prime numbers, much more. Bibliography. 380pp. 5⅜ x 8½. 65620-9 Pa. $10.95

THE VARIATIONAL PRINCIPLES OF MECHANICS, Cornelius Lanczos. Graduate level coverage of calculus of variations, equations of motion, relativistic mechanics, more. First inexpensive paperbound edition of classic treatise. Index. Bibliography. 418pp. 5⅜ x 8½. 65067-7 Pa. $12.95

MATHEMATICAL TABLES AND FORMULAS, Robert D. Carmichael and Edwin R. Smith. Logarithms, sines, tangents, trig functions, powers, roots, reciprocals, exponential and hyperbolic functions, formulas and theorems. 269pp. 5⅜ x 8½. 60111-0 Pa. $6.95

THEORETICAL PHYSICS, Georg Joos, with Ira M. Freeman. Classic overview covers essential math, mechanics, electromagnetic theory, thermodynamics, quantum mechanics, nuclear physics, other topics. First paperback edition. xxiii + 885pp. 5⅜ x 8½. 65227-0 Pa. $21.95

HANDBOOK OF MATHEMATICAL FUNCTIONS WITH FORMULAS, GRAPHS, AND MATHEMATICAL TABLES, edited by Milton Abramowitz and Irene A. Stegun. Vast compendium: 29 sets of tables, some to as high as 20 places. 1,046pp. 8 x 10½. 61272-4 Pa. $26.95

MATHEMATICAL METHODS IN PHYSICS AND ENGINEERING, John W. Dettman. Algebraically based approach to vectors, mapping, diffraction, other topics in applied math. Also generalized functions, analytic function theory, more. Exercises. 448pp. 5⅜ x 8¼. 65649-7 Pa. $10.95

A SURVEY OF NUMERICAL MATHEMATICS, David M. Young and Robert Todd Gregory. Broad self-contained coverage of computer-oriented numerical algorithms for solving various types of mathematical problems in linear algebra, ordinary and partial, differential equations, much more. Exercises. Total of 1,248pp. 5⅜ x 8½. Two volumes. Vol. I: 65691-8 Pa. $16.95
Vol. II: 65692-6 Pa. $16.95

TENSOR ANALYSIS FOR PHYSICISTS, J.A. Schouten. Concise exposition of the mathematical basis of tensor analysis, integrated with well-chosen physical examples of the theory. Exercises. Index. Bibliography. 289pp. 5⅜ x 8½. 65582-2 Pa. $8.95

INTRODUCTION TO NUMERICAL ANALYSIS (2nd Edition), F.B. Hildebrand. Classic, fundamental treatment covers computation, approximation, interpolation, numerical differentiation and integration, other topics. 150 new problems. 669pp. 5⅜ x 8½. 65363-3 Pa. $16.95

INVESTIGATIONS ON THE THEORY OF THE BROWNIAN MOVEMENT, Albert Einstein. Five papers (1905–8) investigating dynamics of Brownian motion and evolving elementary theory. Notes by R. Fürth. 122pp. 5⅜ x 8½. 60304-0 Pa. $5.95

CATASTROPHE THEORY FOR SCIENTISTS AND ENGINEERS, Robert Gilmore. Advanced-level treatment describes mathematics of theory grounded in the work of Poincaré, R. Thom, other mathematicians. Also important applications to problems in mathematics, physics, chemistry and engineering. 1981 edition. References. 28 tables. 397 black-and-white illustrations. xvii + 666pp. 6⅛ x 9¼. 67539-4 Pa. $17.95

AN INTRODUCTION TO STATISTICAL THERMODYNAMICS, Terrell L. Hill. Excellent basic text offers wide-ranging coverage of quantum statistical mechanics, systems of interacting molecules, quantum statistics, more. 523pp. 5⅜ x 8½. 65242-4 Pa. $12.95

STATISTICAL PHYSICS, Gregory H. Wannier. Classic text combines thermodynamics, statistical mechanics and kinetic theory in one unified presentation of thermal physics. Problems with solutions. Bibliography. 532pp. 5⅜ x 8½. 65401-X Pa. $12.95

ORDINARY DIFFERENTIAL EQUATIONS, Morris Tenenbaum and Harry Pollard. Exhaustive survey of ordinary differential equations for undergraduates in mathematics, engineering, science. Thorough analysis of theorems. Diagrams. Bibliography. Index. 818pp. 5⅜ x 8½. 64940-7 Pa. $18.95

STATISTICAL MECHANICS: Principles and Applications, Terrell L. Hill. Standard text covers fundamentals of statistical mechanics, applications to fluctuation theory, imperfect gases, distribution functions, more. 448pp. 5⅜ x 8½. 65390-0 Pa. $11.95

ORDINARY DIFFERENTIAL EQUATIONS AND STABILITY THEORY: An Introduction, David A. Sánchez. Brief, modern treatment. Linear equation, stability theory for autonomous and nonautonomous systems, etc. 164pp. 5⅜ x 8¼. 63828-6 Pa. $6.95

THIRTY YEARS THAT SHOOK PHYSICS: The Story of Quantum Theory, George Gamow. Lucid, accessible introduction to influential theory of energy and matter. Careful explanations of Dirac's anti-particles, Bohr's model of the atom, much more. 12 plates. Numerous drawings. 240pp. 5⅜ x 8½. 24895-X Pa. $7.95

THEORY OF MATRICES, Sam Perlis. Outstanding text covering rank, nonsingularity and inverses in connection with the development of canonical matrices under the relation of equivalence, and without the intervention of determinants. Includes exercises. 237pp. 5⅜ x 8½. 66810-X Pa. $8.95

GREAT EXPERIMENTS IN PHYSICS: Firsthand Accounts from Galileo to Einstein, edited by Morris H. Shamos. 25 crucial discoveries: Newton's laws of motion, Chadwick's study of the neutron, Hertz on electromagnetic waves, more. Original accounts clearly annotated. 370pp. 5⅜ x 8½. 25346-5 Pa. $10.95

INTRODUCTION TO PARTIAL DIFFERENTIAL EQUATIONS WITH APPLICATIONS, E.C. Zachmanoglou and Dale W. Thoe. Essentials of partial differential equations applied to common problems in engineering and the physical sciences. Problems and answers. 416pp. 5⅜ x 8½. 65251-3 Pa. $11.95

BURNHAM'S CELESTIAL HANDBOOK, Robert Burnham, Jr. Thorough guide to the stars beyond our solar system. Exhaustive treatment. Alphabetical by constellation: Andromeda to Cetus in Vol. 1; Chamaeleon to Orion in Vol. 2; and Pavo to Vulpecula in Vol. 3. Hundreds of illustrations. Index in Vol. 3. 2,000pp. 6⅛ x 9¼. 23567-X, 23568-8, 23673-0 Pa., Three-vol. set $44.85

CHEMICAL MAGIC, Leonard A. Ford. Second Edition, Revised by E. Winston Grundmeier. Over 100 unusual stunts demonstrating cold fire, dust explosions, much more. Text explains scientific principles and stresses safety precautions. 128pp. 5⅜ x 8½. 67628-5 Pa. $5.95

AMATEUR ASTRONOMER'S HANDBOOK, J.B. Sidgwick. Timeless, comprehensive coverage of telescopes, mirrors, lenses, mountings, telescope drives, micrometers, spectroscopes, more. 189 illustrations. 576pp. 5⅜ x 8¼. (Available in U.S. only) 24034-7 Pa. $11.95

SPECIAL FUNCTIONS, N.N. Lebedev. Translated by Richard Silverman. Famous Russian work treating more important special functions, with applications to specific problems of physics and engineering. 38 figures. 308pp. 5⅜ x 8½. 60624-4 Pa. $9.95

OBSERVATIONAL ASTRONOMY FOR AMATEURS, J.B. Sidgwick. Mine of useful data for observation of sun, moon, planets, asteroids, aurorae, meteors, comets, variables, binaries, etc. 39 illustrations. 384pp. 5⅜ x 8¼. (Available in U.S. only) 24033-9 Pa. $8.95

INTEGRAL EQUATIONS, F.G. Tricomi. Authoritative, well-written treatment of extremely useful mathematical tool with wide applications. Volterra Equations, Fredholm Equations, much more. Advanced undergraduate to graduate level. Exercises. Bibliography. 238pp. 5⅜ x 8½. 64828-1 Pa. $8.95

POPULAR LECTURES ON MATHEMATICAL LOGIC, Hao Wang. Noted logician's lucid treatment of historical developments, set theory, model theory, recursion theory and constructivism, proof theory, more. 3 appendixes. Bibliography. 1981 edition. ix + 283pp. 5⅜ x 8½. 67632-3 Pa. $8.95

MODERN NONLINEAR EQUATIONS, Thomas L. Saaty. Emphasizes practical solution of problems; covers seven types of equations. ". . . a welcome contribution to the existing literature...."–*Math Reviews*. 490pp. 5⅜ x 8½. 64232-1 Pa. $13.95

FUNDAMENTALS OF ASTRODYNAMICS, Roger Bate et al. Modern approach developed by U.S. Air Force Academy. Designed as a first course. Problems, exercises. Numerous illustrations. 455pp. 5⅜ x 8½. 60061-0 Pa. $10.95

INTRODUCTION TO LINEAR ALGEBRA AND DIFFERENTIAL EQUATIONS, John W. Dettman. Excellent text covers complex numbers, determinants, orthonormal bases, Laplace transforms, much more. Exercises with solutions. Undergraduate level. 416pp. 5⅜ x 8½. 65191-6 Pa. $11.95

INCOMPRESSIBLE AERODYNAMICS, edited by Bryan Thwaites. Covers theoretical and experimental treatment of the uniform flow of air and viscous fluids past two-dimensional aerofoils and three-dimensional wings; many other topics. 654pp. 5⅜ x 8½. 65465-6 Pa. $16.95

INTRODUCTION TO DIFFERENCE EQUATIONS, Samuel Goldberg. Exceptionally clear exposition of important discipline with applications to sociology, psychology, economics. Many illustrative examples; over 250 problems. 260pp. 5⅜ x 8½. 65084-7 Pa. $8.95

LAMINAR BOUNDARY LAYERS, edited by L. Rosenhead. Engineering classic covers steady boundary layers in two- and three- dimensional flow, unsteady boundary layers, stability, observational techniques, much more. 708pp. 5⅜ x 8½. 65646-2 Pa. $18 95

LECTURES ON CLASSICAL DIFFERENTIAL GEOMETRY, Second Edition, Dirk J. Struik. Excellent brief introduction covers curves, theory of surfaces, fundamental equations, geometry on a surface, conformal mapping, other topics. Problems. 240pp. 5⅜ x 8½. 65609-8 Pa. $8.95

ROTARY-WING AERODYNAMICS, W.Z. Stepniewski. Clear, concise text covers aerodynamic phenomena of the rotor and offers guidelines for helicopter performance evaluation. Originally prepared for NASA. 537 figures. 640pp. 6⅛ x 9¼.
64647-5 Pa. $16.95

DIFFERENTIAL GEOMETRY, Heinrich W. Guggenheimer. Local differential geometry as an application of advanced calculus and linear algebra. Curvature, transformation groups, surfaces, more. Exercises. 62 figures. 378pp. 5⅜ x 8½.
63433-7 Pa. $9.95

INTRODUCTION TO SPACE DYNAMICS, William Tyrrell Thomson. Comprehensive, classic introduction to space-flight engineering for advanced undergraduate and graduate students. Includes vector algebra, kinematics, transformation of coordinates. Bibliography. Index. 352pp. 5⅜ x 8½.
65113-4 Pa. $9.95

A SURVEY OF MINIMAL SURFACES, Robert Osserman. Up-to-date, in-depth discussion of the field for advanced students. Corrected and enlarged edition covers new developments. Includes numerous problems. 192pp. 5⅜ x 8½. 64998-9 Pa. $8.95

ANALYTICAL MECHANICS OF GEARS, Earle Buckingham. Indispensable reference for modern gear manufacture covers conjugate gear-tooth action, gear-tooth profiles of various gears, many other topics. 263 figures. 102 tables. 546pp. 5⅜ x 8½.
65712-4 Pa. $14.95

SET THEORY AND LOGIC, Robert R. Stoll. Lucid introduction to unified theory of mathematical concepts. Set theory and logic seen as tools for conceptual understanding of real number system. 496pp. 5⅜ x 8¼. 63829-4 Pa. $12.95

A HISTORY OF MECHANICS, René Dugas. Monumental study of mechanical principles from antiquity to quantum mechanics. Contributions of ancient Greeks, Galileo, Leonardo, Kepler, Lagrange, many others. 671pp. 5⅜ x 8½.
65632-2 Pa. $14.95

FAMOUS PROBLEMS OF GEOMETRY AND HOW TO SOLVE THEM, Benjamin Bold. Squaring the circle, trisecting the angle, duplicating the cube: learn their history, why they are impossible to solve, then solve them yourself. 128pp. 5⅜ x 8½. 24297-8 Pa. $4.95

MECHANICAL VIBRATIONS, J.P. Den Hartog. Classic textbook offers lucid explanations and illustrative models, applying theories of vibrations to a variety of practical industrial engineering problems. Numerous figures. 233 problems, solutions. Appendix. Index. Preface. 436pp. 5⅜ x 8½. 64785-4 Pa. $11.95

CURVATURE AND HOMOLOGY, Samuel I. Goldberg. Thorough treatment of specialized branch of differential geometry. Covers Riemannian manifolds, topology of differentiable manifolds, compact Lie groups, other topics. Exercises. 315pp. 5⅜ x 8½. 64314-X Pa. $9.95

HISTORY OF STRENGTH OF MATERIALS, Stephen P. Timoshenko. Excellent historical survey of the strength of materials with many references to the theories of elasticity and structure. 245 figures. 452pp. 5⅜ x 8½. 61187-6 Pa. $12.95

CATALOG OF DOVER BOOKS

GEOMETRY OF COMPLEX NUMBERS, Hans Schwerdtfeger. Illuminating, widely praised book on analytic geometry of circles, the Moebius transformation, and two-dimensional non-Euclidean geometries. 200pp. 5⅜ x 8¼. 63830-8 Pa. $8.95

MECHANICS, J.P. Den Hartog. A classic introductory text or refresher. Hundreds of applications and design problems illuminate fundamentals of trusses, loaded beams and cables, etc. 334 answered problems. 462pp. 5⅜ x 8½. 60754-2 Pa. $11.95

TOPOLOGY, John G. Hocking and Gail S. Young. Superb one-year course in classical topology. Topological spaces and functions, point-set topology, much more. Examples and problems. Bibliography. Index. 384pp. 5⅜ x 8¼. 65676-4 Pa. $10.95

STRENGTH OF MATERIALS, J.P. Den Hartog. Full, clear treatment of basic material (tension, torsion, bending, etc.) plus advanced material on engineering methods, applications. 350 answered problems. 323pp. 5⅜ x 8½. 60755-0 Pa. $9.95

ELEMENTARY CONCEPTS OF TOPOLOGY, Paul Alexandroff. Elegant, intuitive approach to topology from set-theoretic topology to Betti groups; how concepts of topology are useful in math and physics. 25 figures. 57pp. 5⅜ x 8½. 60747-X Pa. $3.95

ADVANCED STRENGTH OF MATERIALS, J.P. Den Hartog. Superbly written advanced text covers torsion, rotating disks, membrane stresses in shells, much more. Many problems and answers. 388pp. 5⅜ x 8½. 65407-9 Pa. $10.95

COMPUTABILITY AND UNSOLVABILITY, Martin Davis. Classic graduate-level introduction to theory of computability, usually referred to as theory of recurrent functions. New preface and appendix. 288pp. 5⅜ x 8½. 61471-9 Pa. $8.95

GENERAL CHEMISTRY, Linus Pauling. Revised 3rd edition of classic first-year text by Nobel laureate. Atomic and molecular structure, quantum mechanics, statistical mechanics, thermodynamics correlated with descriptive chemistry. Problems. 992pp. 5⅜ x 8½. 65622-5 Pa. $19.95

AN INTRODUCTION TO MATRICES, SETS AND GROUPS FOR SCIENCE STUDENTS, G. Stephenson. Concise, readable text introduces sets, groups, and most importantly, matrices to undergraduate students of physics, chemistry, and engineering. Problems. 164pp. 5⅜ x 8½. 65077-4 Pa. $7.95

THE HISTORICAL BACKGROUND OF CHEMISTRY, Henry M. Leicester. Evolution of ideas, not individual biography. Concentrates on formulation of a coherent set of chemical laws. 260pp. 5⅜ x 8½. 61053-5 Pa. $8.95

THE PHILOSOPHY OF MATHEMATICS: An Introductory Essay, Stephan Körner. Surveys the views of Plato, Aristotle, Leibniz & Kant concerning propositions and theories of applied and pure mathematics. Introduction. Two appendices. Index. 198pp. 5⅜ x 8½. 25048-2 Pa. $8.95

THE DEVELOPMENT OF MODERN CHEMISTRY, Aaron J. Ihde. Authoritative history of chemistry from ancient Greek theory to 20th-century innovation. Covers major chemists and their discoveries. 209 illustrations. 14 tables. Bibliographies. Indices. Appendices. 851pp. 5⅜ x 8½. 64235-6 Pa. $18.95

DE RE METALLICA, Georgius Agricola. The famous Hoover translation of greatest treatise on technological chemistry, engineering, geology, mining of early modern times (1556). All 289 original woodcuts. 638pp. 6¾ x 11. 60006-8 Pa. $21.95

SOME THEORY OF SAMPLING, William Edwards Deming. Analysis of the problems, theory and design of sampling techniques for social scientists, industrial managers and others who find statistics increasingly important in their work. 61 tables. 90 figures. xvii + 602pp. 5⅜ x 8½. 64684-X Pa. $16.95

THE VARIOUS AND INGENIOUS MACHINES OF AGOSTINO RAMELLI: A Classic Sixteenth-Century Illustrated Treatise on Technology, Agostino Ramelli. One of the most widely known and copied works on machinery in the 16th century. 194 detailed plates of water pumps, grain mills, cranes, more. 608pp. 9 x 12.
28180-9 Pa. $24.95

LINEAR PROGRAMMING AND ECONOMIC ANALYSIS, Robert Dorfman, Paul A. Samuelson and Robert M. Solow. First comprehensive treatment of linear programming in standard economic analysis. Game theory, modern welfare economics, Leontief input-output, more. 525pp. 5⅜ x 8½. 65491-5 Pa. $14.95

ELEMENTARY DECISION THEORY, Herman Chernoff and Lincoln E. Moses. Clear introduction to statistics and statistical theory covers data processing, probability and random variables, testing hypotheses, much more. Exercises. 364pp. 5⅜ x 8½. 65218-1 Pa. $10.95

THE COMPLEAT STRATEGYST: Being a Primer on the Theory of Games of Strategy, J.D. Williams. Highly entertaining classic describes, with many illustrated examples, how to select best strategies in conflict situations. Prefaces. Appendices. 268pp. 5⅜ x 8½. 25101-2 Pa. $7.95

CONSTRUCTIONS AND COMBINATORIAL PROBLEMS IN DESIGN OF EXPERIMENTS, Damaraju Raghavarao. In-depth reference work examines orthogonal Latin squares, incomplete block designs, tactical configuration, partial geometry, much more. Abundant explanations, examples. 416pp. 5⅜ x 8¼.
65685-3 Pa. $10.95

THE ABSOLUTE DIFFERENTIAL CALCULUS (CALCULUS OF TENSORS), Tullio Levi-Civita. Great 20th-century mathematician's classic work on material necessary for mathematical grasp of theory of relativity. 452pp. 5⅜ x 8½.
63401-9 Pa. $11.95

VECTOR AND TENSOR ANALYSIS WITH APPLICATIONS, A.I. Borisenko and I.E. Tarapov. Concise introduction. Worked-out problems, solutions, exercises. 257pp. 5⅝ x 8¼. 63833-2 Pa. $8.95

THE FOUR-COLOR PROBLEM: Assaults and Conquest, Thomas L. Saaty and Paul G. Kainen. Engrossing, comprehensive account of the century-old combinatorial topological problem, its history and solution. Bibliographies. Index. 110 figures. 228pp. 5⅜ x 8½. 65092-8 Pa. $7.95

CATALYSIS IN CHEMISTRY AND ENZYMOLOGY, William P. Jencks. Exceptionally clear coverage of mechanisms for catalysis, forces in aqueous solution, carbonyl- and acyl-group reactions, practical kinetics, more. 864pp. 5⅜ x 8½.
65460-5 Pa. $19.95

PROBABILITY: An Introduction, Samuel Goldberg. Excellent basic text covers set theory, probability theory for finite sample spaces, binomial theorem, much more. 360 problems. Bibliographies. 322pp. 5⅜ x 8½.
65252-1 Pa. $10.95

LIGHTNING, Martin A. Uman. Revised, updated edition of classic work on the physics of lightning. Phenomena, terminology, measurement, photography, spectroscopy, thunder, more. Reviews recent research. Bibliography. Indices. 320pp. 5⅜ x 8¼.
64575-4 Pa. $8.95

PROBABILITY THEORY: A Concise Course, Y.A. Rozanov. Highly readable, self-contained introduction covers combination of events, dependent events, Bernoulli trials, etc. Translation by Richard Silverman. 148pp. 5⅜ x 8¼.
63544-9 Pa. $7.95

AN INTRODUCTION TO HAMILTONIAN OPTICS, H. A. Buchdahl. Detailed account of the Hamiltonian treatment of aberration theory in geometrical optics. Many classes of optical systems defined in terms of the symmetries they possess. Problems with detailed solutions. 1970 edition. xv + 360pp. 5⅜ x 8½.
67597-1 Pa. $10.95

STATISTICS MANUAL, Edwin L. Crow, et al. Comprehensive, practical collection of classical and modern methods prepared by U.S. Naval Ordnance Test Station. Stress on use. Basics of statistics assumed. 288pp. 5⅜ x 8½.
60599-X Pa. $7.95

DICTIONARY/OUTLINE OF BASIC STATISTICS, John E. Freund and Frank J. Williams. A clear concise dictionary of over 1,000 statistical terms and an outline of statistical formulas covering probability, nonparametric tests, much more. 208pp. 5⅜ x 8½.
66796-0 Pa. $7.95

STATISTICAL METHOD FROM THE VIEWPOINT OF QUALITY CONTROL, Walter A. Shewhart. Important text explains regulation of variables, uses of statistical control to achieve quality control in industry, agriculture, other areas. 192pp. 5⅜ x 8½.
65232-7 Pa. $7.95

METHODS OF THERMODYNAMICS, Howard Reiss. Outstanding text focuses on physical technique of thermodynamics, typical problem areas of understanding, and significance and use of thermodynamic potential. 1965 edition. 238pp. 5⅜ x 8½.
69445-3 Pa. $8.95

STATISTICAL ADJUSTMENT OF DATA, W. Edwards Deming. Introduction to basic concepts of statistics, curve fitting, least squares solution, conditions without parameter, conditions containing parameters. 26 exercises worked out. 271pp. 5⅜ x 8½.
64685-8 Pa. $9.95

TENSOR CALCULUS, J.L. Synge and A. Schild. Widely used introductory text covers spaces and tensors, basic operations in Riemannian space, non-Riemannian spaces, etc. 324pp. 5⅜ x 8¼.
63612-7 Pa. $9.95

A CONCISE HISTORY OF MATHEMATICS, Dirk J. Struik. The best brief history of mathematics. Stresses origins and covers every major figure from ancient Near East to 19th century. 41 illustrations. 195pp. 5⅜ x 8½. 60255-9 Pa. $8.95

A SHORT ACCOUNT OF THE HISTORY OF MATHEMATICS, W.W. Rouse Ball. One of clearest, most authoritative surveys from the Egyptians and Phoenicians through 19th-century figures such as Grassman, Galois, Riemann. Fourth edition. 522pp. 5⅜ x 8½. 20630-0 Pa. $11.95

HISTORY OF MATHEMATICS, David E. Smith. Nontechnical survey from ancient Greece and Orient to late 19th century; evolution of arithmetic, geometry, trigonometry, calculating devices, algebra, the calculus. 362 illustrations. 1,355pp. 5⅜ x 8½. 20429-4, 20430-8 Pa., Two-vol. set $26.90

THE GEOMETRY OF RENÉ DESCARTES, René Descartes. The great work founded analytical geometry. Original French text, Descartes' own diagrams, together with definitive Smith-Latham translation. 244pp. 5⅜ x 8½. 60068-8 Pa. $8.95

THE ORIGINS OF THE INFINITESIMAL CALCULUS, Margaret E. Baron. Only fully detailed and documented account of crucial discipline: origins; development by Galileo, Kepler, Cavalieri; contributions of Newton, Leibniz, more. 304pp. 5⅜ x 8½. (Available in U.S. and Canada only) 65371-4 Pa. $9.95

THE HISTORY OF THE CALCULUS AND ITS CONCEPTUAL DEVELOPMENT, Carl B. Boyer. Origins in antiquity, medieval contributions, work of Newton, Leibniz, rigorous formulation. Treatment is verbal. 346pp. 5⅜ x 8½. 60509-4 Pa. $9.95

THE THIRTEEN BOOKS OF EUCLID'S ELEMENTS, translated with introduction and commentary by Sir Thomas L. Heath. Definitive edition. Textual and linguistic notes, mathematical analysis. 2,500 years of critical commentary. Not abridged. 1,414pp. 5⅜ x 8½. 60088-2, 60089-0, 60090-4 Pa., Three-vol. set $32.85

GAMES AND DECISIONS: Introduction and Critical Survey, R. Duncan Luce and Howard Raiffa. Superb nontechnical introduction to game theory, primarily applied to social sciences. Utility theory, zero-sum games, n-person games, decision-making, much more. Bibliography. 509pp. 5⅜ x 8½. 65943-7 Pa. $13.95

THE HISTORICAL ROOTS OF ELEMENTARY MATHEMATICS, Lucas N.H. Bunt, Phillip S. Jones, and Jack D. Bedient. Fundamental underpinnings of modern arithmetic, algebra, geometry and number systems derived from ancient civilizations. 320pp. 5⅜ x 8½. 25563-8 Pa. $8.95

CALCULUS REFRESHER FOR TECHNICAL PEOPLE, A. Albert Klaf. Covers important aspects of integral and differential calculus via 756 questions. 566 problems, most answered. 431pp. 5⅜ x 8½. 20370-0 Pa. $8.95